Nanoscale Quantum Materials

Nanoscale Quantum Materials

Musings on the Ultra-Small World

Tapash Chakraborty

CRC Press
Taylor & Francis Group
Boca Raton London New York

CRC Press is an imprint of the
Taylor & Francis Group, an **informa** business

First edition published [2022]
by CRC Press
6000 Broken Sound Parkway NW, Suite 300, Boca Raton, FL 33487-2742

and by CRC Press
2 Park Square, Milton Park, Abingdon, Oxon, OX14 4RN

Library of Congress Cataloging-in-Publication Data
Names: Chakraborty, T. (Tapash), 1950- author.
Title: Nanoscale quantum materials : musings on the ultra-small world / Tapash Chakraborty.
Description: First edition. | Boca Raton : CRC Press, 2022. | Includes bibliographical references and index.
Identifiers: LCCN 2021023346 | ISBN 9780367546397 (hardback) | ISBN 9780367548605 (paperback) | ISBN 9780367548605 (ebook)
Subjects: LCSH: Nanostructured materials. | Nanotechnology. | Quantum dots.
Classification: LCC TA418.9.N35 C3685 2022 | DDC 620.1/15--dc23
LC record available at https://lccn.loc.gov/2021023346

ISBN: 978-0-367-54639-7 (hbk)
ISBN: 978-0-367-54860-5 (pbk)
ISBN: 978-1-003-09090-8 (ebk)

DOI: 10.1201/9781003090908

Typeset in LM Roman
by KnowledgeWorks Global Ltd.

Contents

Preface ix

Acknowledgments xv

1 Introduction: From giants to dwarfs 1

2 Down to low dimensions 9
   2.1 The essential toolkit: Quantum mechanics 9
   2.2 Two-dimensional electron gas 10
   2.3 Novel phenomena in flatland: Nobels galore 15
       2.3.1 Quantum Hall effect 16
       2.3.2 A new standard for resistance calibration 20
       2.3.3 Quantum Hall effect – now with fractions 22
   2.4 Laughlin's eponymous wave function 24
       2.4.1 The (lowest energy) ground state 25
   2.5 Incompressible liquid and the charged excitations 30
   2.6 The unusual statistics 33
   2.7 Spin flip and the tilted magnetic field 36
   2.8 Anatomy of the Laughlin state 39
   2.9 Laughlin state from the East 43

3 Quantum dots: In the abyss of no dimensions 47
   3.1 Landau versus Fock 49
   3.2 A tale of artificial atoms 52
   3.3 Portrait of a harmonic oscillator 53
   3.4 Magic number ground states 56
   3.5 Rashba spin-orbit coupling 58
   3.6 Spin textures and topological charge 64
   3.7 Anisotropic quantum dots 68
   3.8 Secret affairs and a single photon 70
       3.8.1 Single-photon detectors 73
       3.8.2 Single-photon source 75

3.9   Cascading and burning bright   77

        3.9.1   Molecular fingerprinting   77

        3.9.2   Quantum cascade laser   78

        3.9.3   QCL in a magnetic field   79

        3.9.4   QCL with quantum dots   83

4   Quantum rings: Dynamic unity of polar opposites   87

4.1   Tireless electron running around in circles   88

4.2   Interacting electrons in a few-electron quantum ring   . .   91

4.3   Optical spectroscopy   94

4.4   Role of electron spin   100

4.5   Quantum ring complexes   104

4.6   Rashba spin-orbit coupling revisited   105

4.7   Quantum ring and topological charge   107

4.8   Rings in novel systems   110

4.9   Isotropic or anisotropic?   115

4.10   Device applications of the quantum rings   116

5   Graphene: Carbon and its nets   119

5.1   A brief history of graphene   120

        5.1.1   Major breakthroughs in graphene research   121

        5.1.2   Isolating graphene: Sellotape versus the Scotch
                tape   125

5.2   Electrons behaving differently   127

5.3   Quantum Hall effects in graphene   129

5.4   Bilayer graphene   132

        5.4.1   Bilayer graphene Landau levels   135

        5.4.2   Novel fractional quantum Hall effects   136

        5.4.3   Bilayer graphene in a tilted magnetic field   144

        5.4.4   Marvels of interacting electrons in graphene . . .   147

5.5   Graphene nanostructures   148

        5.5.1   Quantum dots   148

        5.5.2   Quantum rings   150

5.6   Molecular adsorption on graphene   156

5.7   Graphene's extended family   158

5.8   Requiem for the (very expensive) dreams   160

6   Some remarkable episodes in the nanoscale   165

6.1   Fractal butterflies   165

        6.1.1   Semiconductor systems: Strong field limit   169

6.1.2 Butterflies in monolayer graphene 171
6.1.3 Square lattice periodic structure 173
6.1.4 Moiré structure 176
6.1.5 Butterflies in bilayer graphene 178
6.1.6 Butterflies and interacting electrons 181
6.1.7 The Cantor set 188
6.1.8 The ten-martini challenge 190
6.2 Maxwell's demon in the nanoworld 193
6.2.1 A sorting demon – the anti-thermodynamic agent 195
6.2.2 Anthropomorphism of the benevolent demon . . . 197
6.2.3 Demon in quantum dots 199
6.2.4 Spin demons in quantum rings 201
6.3 Nanoscale physics of DNA 205
6.3.1 DNA – Nature's nanoscale code-script 205
6.3.2 DNA electronics 209
6.3.3 Humidity assisted conduction 210
6.3.4 Mismatched base pairs: Electrical properties . . . 212

7 Epilogue and the road ahead   **217**

A Ten-martini proof   **221**

Index   **229**

Preface

Our everyday life today is more and more enmeshed with smart electronic devices that, in their core, contain nanoscale objects. For researchers, a deeper knowledge of the novel properties of those objects is absolutely essential to make them perform ever more efficiently. Intense research on physics of nanoscale systems has uncovered many unexpected phenomena that has helped push forward the frontiers of our present understanding of Nature. Sharing that knowledge with the general reader, who is not well versed about those complicated phenomena, is important and essential for our collective appreciation of the rapid developments of this digital age that has been dominating our daily lives. In this book, I have endeavored to achieve that to some extent, without claiming the elevated clarity of a popular science book or an authoritative overview of a textbook. The essential physics at this level is rather complex, but my goal is to explain the basic principles of the various phenomena as plainly as I can without resorting to complicated equations. However, for more technical-minded readers, I have presented some technical-details in boxes or footnotes that can be skipped for a general reading of the book.

An alert reader will surely notice that while writing this book I have taken some poetic license to describe some of the topics in order to make those more palatable to a wider audience. My fervent hope has been that it will appeal not only to physics researchers, engineers, and advanced students in those disciplines, but perhaps also to 'physics enthusiasts' from other academic disciplines. However, some familiarities with the fundamentals of quantum mechanics and condensed matter physics will be essential to appreciate the book in its entirety.

The book is by no means a comprehensive treatise of all things nanoscale. It is also biased toward the theoretical perspectives of nanoscale physics[1], although the relevant experimental works associated with those ideas will also be discussed. Theoretical research in physics always had profound impact on the course of research that we expound in this book. Since the founding of theoretical physics by none other than Galileo Galilei, it has made physics 'the most successful of all the sciences in the description of Nature'[2]. Indeed the beauty and strength of theoretical physics lie in its ability to unravel the laws of Nature by pure intellect[3]. One great example in this respect would be, of course, Maxwell's equations[4], which are considered to be among the most influential in the history of science[5]. Reported in the Victorian era, they are still essential to describing how magnets work, explain the marvels of electricity, and most importantly, describe the nature of light that leads us out of darkness and is vital to sustain life on this planet. Maxwell's equations are used regularly in physics, engineering, and as disparate a field as neurophysiology, and are still very relevant for today's research of nanoscience! In this book, we will revisit one of Maxwell's other timeless creations, the intelligent demon (see Chapter 6).

A couple of centuries later, pioneers such as Max Planck, Erwin Schrödinger, Werner Heisenberg, and others laid the foundations of quantum mechanics that has made possible nanoscale research which has been impacting our life so profoundly today[6]! In our era, Bob Laughlin's introduction of the eponymous wave function in 1983 to explain the fractional quantum Hall effect has been one such milestone. He 'set the stage for one of the most beautiful developments in the physics of the

---

[1]For a comprehensive treatment of nanoscience from an experimentalists' point of view, please see, e.g., *Quantum Materials*, edited by D. Heitmann (Springer, 2010); *Nanoscience: The Science of the Small in Physics, Engineering, Chemistry, Biology, and Medicine* by Hans-Eckhardt Schaefer (Springer, Heidelberg 2010).

[2]Galileo Galilei, 1564-1642, and The motion of falling bodies, by R.B. Lindsay, *Amer. J. Phys.* **10**, 285 (1942).

[3]*The Beautiful Invisible: Creativity, Imagination, and Theoretical Physics*, by Giovanni Vignale, Oxford University Press, 2011).

[4]A set of four most fundamental equations in all of science, which constitute a *complete* description of the classical behavior of electric and magnetic fields. These equations provided a theoretical description of light as an electromagnetic wave of very short wavelength. Maxwell: A new vision of the world, by D. Maystre,*C.R. Physique* **15**, 387 (2014).

[5]*The Digital Mind: How Science is Redefining Humanity*, by Arlindo Oliveira (MIT Press, 2017).

[6]*Quantum Physics in the Nanoworld*, by Hans Lüth (Springer, Heidelberg 2009).

twentieth century'[7], by introducing a new type of quantum fluid with several unique and unanticipated properties that has fascinated thousands of researchers and triggered an avalanche of novel ideas yet to subside. Its impact has even been felt in many other branches of physics.

Unlike in olden days, 'novel' ideas in physics are now reported on a daily basis. The publishers are keen to publish more and more groundbreaking results whose shelf life is obviously limited. In this situation, I believe sometimes it is perhaps very refreshing to step back from those cacophonies of drum beats from the journals and the authors themselves about their latest presumptive earth-shattering discoveries, and try to look in more traditional ways the real progress that has been made in our field. But why even attempt to write a less technical book on such highly specialized topics given that there are so many technical books or review articles containing many of the topics described in this book are already available and are perhaps not as challenging to write (if you have remained active for years in that field). However, it is always desirable (and certainly rewarding) to make an effort to explain all the highly complex topics to non-experts. The important question is, of course, how well can it be done?

The limitations of this pursuit are well known and keenly felt while choosing the topics for this book. The ineffability between, for example, what quantum mechanics says and what the words say is so vast that it has bothered thinkers for decades. As the literary scholar George Steiner wrote in *The Retreat from the Word* about the 'unspeakability' of modern science[8]: 'It is during the seventeenth century that significant areas of truth, reality, and action recede from the sphere of verbal statement'. 'The great book of the Universe', wrote Galileo, 'is written in mathematical language'[9].

---

[7] *Quantum Theory of the Electron Liquid*, by G.F. Giuliani and G. Vignale, Cambridge University Press, 2005.

[8] *The Kenyon Review* **23**, 187 (1961).

[9] Il saggiatore, by G. Galilei (1623), p. 11. The actual text was 'La filosofia è scritta in questo grandissimo libro che continuamente ci sta aperto innanzi a gli occhi (io dico l'universo), ma non si può intendere se prima non s'impara a intender la lingua, e conoscer i caratteri, ne' quali è scritto. Egli è scritto in lingua matematica, e i caratteri son triangoli, cerchi, ed altre figure geometriche, senza i quali mezi è impossibile a intenderne umanamente parola; senza questi è un aggirarsi vanamente per un oscuro laberinto'. The English translation by S. Drake, *Discoveries and Opinions of Galileo*, Anchor Books, 1957, goes as follows: Philosophy is written in this grand book, the universe, which stands continually open to our gaze. But the book cannot be understood unless one first learns to comprehend the language and read

In this Galileo was not alone. As a matter of fact, over the past three or four centuries, 'physicists'[10] have begun to rely more on mathematics rather than the word to describe the world. This trend has led to physics, just as mathematics itself, making a retreat from the word. Steiner warned that, 'It is arrogant, if not irresponsible, to invoke such basic notions in our present model of the universe as quanta, the indeterminacy principle, ..., if one cannot do so in the language appropriate to them – that is to say, in mathematical terms. Without it, such words are phantasms to deck out the pretense of philosophers or journalists'. In a somewhat less articulate manner, Ernest Rutherford was quoted as saying if you cannot explain your physics to a barmaid it is probably not very good physics[11]. That must be a daunting job today for any physicist who would embark upon such undertakings. It will be a genuine shock for this author if indeed a barmaid or a poet appreciates the contents of this book!

Interestingly, a few years ago, I interacted with a scientist who considers herself 'an occasional artist and poet' and wrote a very interesting book on the properties of water[12] aimed at readers 'in the arts and humanities as well as scientists'. It has been a motivational factor behind my present quixotic endeavor. 'Here is a scientist who can really write,

---

the letters in which it is composed. It is written in the language of mathematics, and its characters are triangles, circles, and other geometric figures without which it is humanly impossible to understand a single word of it; without these, one wanders about in a dark labyrinth.

[10]We should be careful here about the terminology. The word *physicist* was first proposed by William Whewell in his book, *The Philosophy of the Inductive Sciences*, London, 1840, although those scientists existed for centuries, albeit belonging to a different category! Incidentally, the term *scientist* was also proposed in that book by Whewell as 'a cultivator of science' (see *Science and Technology in World History, An Introduction*, J.E. McClellan III and H. Dorn (Johns Hopkins University Press, Baltimore 2015)). However, according to Sydney Ross, in Scientist: The story of a word, *Annals of Science* **18**, 65 (1964), Whewell introduced the term *scientist* much earlier, somewhat 'jocularly' in *The Quarterly Review* **51**, 58 (1834), in his attempt to find a suitable English equivalent of the German term *natur-forscher*. The undignified compounds, such as *nature-poker* or *nature-peeper*, were 'indignantly rejected' in favor of the word scientist! However, according to Ross, the word was not an instant hit. In fact, for the natural philosophers, 'the word scientist implied making a business of science; it degraded their labors of love to a drudgery for profits or salary'.

[11]Some recollections and reflections on Rutherford, by W. Bennett Lewis, *Notes Rec. R. Soc. Lond.* **27**, 61 (1972).

[12]*Living Rainbow H_2O*, by Mae-Wan Ho (World Scientific 2012).

exulted the legendary Nobelist Hans Bethe[13] some years ago, while reviewing a largely autobiographical book by Freeman Dyson[14]. However, it would be a very rare occasion to witness such a distinction lavished upon a physics author when the narrative involves advanced quantum mechanics.

The book is organized as follows: In Chapter 1, we introduce the subject of scaling from a historical perspective, crediting Galileo for his original contributions in describing the subject of scaling. We explain why scaling down to ever greater miniaturization of electronic circuits and memory devices has enormously improved computer performance and all related electronic devices that, in turn, has resulted in today's all-encompassing digital transformation of our everyday lives. The purpose of this chapter is to prepare the readers for what comes next in the nanoscale regime.

Chapter 2 begins with the description of the two-dimensional electron gas. We explain how the planar electrons can be ideally created, and then we briefly explain how those were actually made in the laboratories. We then highlight the momentous discoveries in those systems and how those effects were explained by introducing novel quantum phenomena that have far reaching implications. We discuss the discovery of the integer quantum Hall effect, the fractional quantum Hall effect, and Laughlin's many-electron wave function for a very special quantum fluid. The fractionally charged quasiparticles and their exotic statistics and various spin configurations are discussed. Every attempt has been made to explain those in such a way that it will be of interest to the non-experts. We have also included several original references and brief technical steps in boxes that might help in grasping a more detailed picture if required.

In Chapter 3, we discuss the properties of zero-dimensional electron systems, the quantum dots. We discuss the fundamental physics of these systems and their possible applications in quantum cryptography and as a source for mid-infrared and terahertz radiation. Chapter 4 deals with the persistent current in small metal rings. The role of interacting electrons in semiconductor quantum rings containing only a few electrons are discussed. Unusual properties of quantum rings in some novel materials and the usefulness of the rings are briefly discussed.

---

[13] *Physics Today*, p. 51, December 1979.

[14] *Disturbing the Universe*, by F. Dyson, Harper & Row, N.Y. 1979.

Chapter 5 deals with the physics of graphene, the so-called 'wonder material' of our time. We briefly trace the historical breakthroughs that made it possible to explore the novel phenomena exhibited in this material. We explain the presence of Dirac fermions in this system with different quantum Hall features seen in monolayer and bilayer graphene. We also explain the nanostructures, such as quantum dots and quantum rings created from graphene. We then briefly explain how the molecular adsorption on graphene can open a band gap absent in pristine graphene. In search of other graphene-like materials, researchers have stumbled upon various other systems with similar, or even better, properties than graphene. We briefly conclude this chapter by explaining how the extreme hype about graphene has not been materialized in real life, notwithstanding the huge finding and manpower being spent chasing perhaps the unattainable.

In the final chapter, we describe a few novel phenomena that are unique to the nanoscale world. We begin with the fractal butterflies, a glimpse of which has finally been found in graphene systems. We briefly discuss how the mathematicians presented the 'ten-martini challenge' to prove the presence of Cantor set in the basic equations of fractal butterflies and how the ten-martini problem was finally mathematically solved. (Here is a topic that perhaps could be of potential interest to our proverbial barmaid!) We then discuss the story of the intelligent demon that Maxwell introduced to understand the challenges posed by the second law of thermodynamics, and how in the nanoscale the demon is making a dramatic comeback. Finally, we briefly introduce DNA, the molecule of life, and explain how it can be incorporated into nanoscale electronics. We also touch upon the DNA damages and the electrical properties of DNA mispairs. We conclude with a look at the possible directions the nanoscale world will hopefully turn to in the near future. A brief sketch of the mathematical proof of the ten-martini problem has been presented in the Appendix.

St. Catharines   *Tapash Chakraborty*
Ontario, Canada

Acknowledgments

My sincere thanks to colleagues and friends, notably Daniel Braak and Peter Maksym, for carefully reading through and commenting on the manuscript. Their invaluable suggestions and criticisms in questions of form and substance greatly improved the exposition and overall organization of the book. The book would have never reached its present form without the scrupulous attention to details of many of my colleagues, in particular, Aram Manaselyan, Wenchen Luo, and Vadim Apalkov, whose valuable input greatly helped improve some of the chapters. My sincere thanks to Dr. Svetlana Jitomirskaya for providing me with the materials presented in the Appendix.

In order to limit the number of pages in the book and the amount of time I would have liked to devote to this project, I have had to make a selection of work that I have discussed here. No attempt has been made to cover every conceivable topic in the field. Hence, it is entirely possible that the reader's favorite work has not been discussed. I apologize in advance for that. The field of low-dimensional electron systems is simply too vast to cover even a fraction of in a book whose focus is rather different.

The invaluable help provided by Hong-Yi Chen from National Taiwan Normal University, Taipei, who has patiently created most of the figures presented in this book, made all the difference in the overall quality of the book. I cannot thank him enough for his time and interest. The cover of the book is graced by a painting (Blue Over Saskatoon, Winter, 2016) by the Canadian artist Marie Lannoo. My heartfelt thanks to Marie for allowing me to reproduce it here. Last, but not least, my thanks to Karine Van Wetering for her valuable gift, a copy of Galileo's book *Two New Sciences*, that fascinated me to delve into the life of this great physicist.

1

Introduction: From giants to dwarfs

We begin our journey to the nanoworld with, perhaps one of the greatest books written in science: *Discorsi E Dimonstrazioni Matematiche, intorno à due nuove Scienze* by Galileo Galilei (published in 1638)[1]. The book is a scientific masterpiece based on his lifelong work, whose one of the basic questions 'what holds solids together' is at the heart of today's condensed matter physics[2]. The virtuosity and fortitude that went into creating this masterpiece is simply mind-blowing, because it is actually difficult to imagine the predicament of the author when this book was published: Blind at the age of 72 due to cataracts and glaucoma, very near to his death (in 1642), under house arrest in his villa in Arcetri, near Florence, by the Roman Inquisition which ruled Galileo's espousing heliocentrism heretic. This all happened after his earlier publication of another magnum opus, the 'Dialogue'[3], which took place among the Aristotelian named Simplicio, a Venetian gentleman Giovanni Sagredo (actually Galileo's close friend of the old days) who represents the educated public and is keen to seek truth, and a scientifically informed Florentine, Filippo Salviati, who was in fact, Galileo's mouthpiece. The subject of the dialogue was the two world systems, that of Ptolemy/Aristotle and that of Copernicus (Fig. 1.1). It was published in Florence in 1632.

Much has been written about Galileo's troubles with the Society of Jesus (the 'Jesuits'), the Roman College, and Pope Urban VIII. Initially, they were all very friendly to Galileo and the Pope even encouraged Galileo to publish the 'Dialogue'. Back then, the Jesuits were, of course, very learned in mathematics, astronomy, except occasionally having a

---

[1] *Dialogues Concerning Two New Sciences*, Translated by Henry Crew and Alfonso de Salvio (Dover, N.Y. 1954). Originally published in 1904 by the MacMillan Company.

[2] Galileo comes to the surface!, by Jeff T.M. De Hosson and A. Cavaleiro, in Cavaleiro A., De Hosson J.T.M. (eds) *Nanostructured Coatings. Nanostructure Science and Technology*, Springer, New York, 2006.

[3] *Dialogue Concerning the Two Chief World Systems*, by Galileo Galilei, Translated by Stillman Drake, The Modern Library, New York, 2001.

DOI: 10.1201/9781003090908-1

FIGURE 1.1
The portrait of three great men of astronomy engaged in conversation, from the engraved frontispiece of Galileo's 'Dialogo', created by the engraver Stephano Della Bella in 1632. From left to right: Aristotle (384-322 BCE), Ptolemy (90-168 CE), and Copernicus (1473-1543). Copernicus is holding a heliocentric model in his left hand. There are also fortifications, guns and a galley visible in the background. Reprinted from *Galileo Galilei: First Physicist*, by J. MacLachlan (Oxford, 1997), with permission from the Oxford University Press.

few oddities such as banning infinitesimals in Italy[4] and Galileo was keen to get their approval. It was certainly one of the motivations behind Galileo's move from Padua to Florence in 1610. However, things turned sour after the book was published, especially when the Pope felt that he was portrayed a simpleton in the book. Although Galileo claimed in the preface that the choice of the name 'simplicio' was in honor of the philosopher Simplicius, the damage was already done, and Galileo was called to stand trial in Rome where he was threatened with torture and forced to 'abjure, curse and detest' his heretical view. His book was placed on the infamous Index librorum prohibitorum, the index of banned books and eventually he was placed under house arrest ('formal imprisonment at the pleasure of the Inquisition') for the rest of his life.

---

[4] *Infinitesimal*, by Amir Alexander (Oneworld Publications, London, 2014).

There are plenty of excellent sources to read about the conflict of Galileo with the Catholic Church (the Galileo affair)[5] and punishment by the Inquisition[6] and will not be elaborated here. The only point worth mentioning is that in the course of propounding the heliocentric universe, the Dialogue contains several brilliant ideas that laid the foundations of modern science. Those include, among others, Galileo's introduction of the law of inertia[7] which was called 'one of the deepest insights in the history of thought'[8] and also the concept of relativity[9] (the Galilean invariance). See also the Footnote[10]. Interestingly it has been claimed that the Medieval Italian poet and philosopher Dante Alighieri in 'The Divine Comedy' already had the idea of relativity some 300 years earlier[11].

Back to our classic book, *Two New Sciences*, after the humiliating defeat at the hands of Urban VIII, Galileo, having no reason to be contrite, simply ploughs on and completes this book, which is largely a testament to his lifetime of scientific work. The same three interlocutors grace the pages of this book as well, discussing in Venice[12] for four days about various ground-breaking and revolutionary ideas in physics and mechanics. Just as in the 'Dialogue', Galileo's representative Salviati is the principal speaker and is the expositor of all the original ideas. Between the other two persons, Sagredo is the more learned and occasionally makes important contributions. The interested layperson Simplicio, raises objections that necessitates alternative interpretations. However, unlike in the Dialogue, they are less antagonistic in this book and even Simplicio is now much more flexible.

---

[5] *The Crime of Galileo*, by Giorgio de Santillana, University of Chicago Press, 1976, *Galileo and the Scientific Revolution*, by Laura Fermi and Gilberto Bernardini, Dover N.Y., 2003, *The Essential Galileo*, by M.A. Finocchiaro, Hackett Publishing Co., Indianapolis, 2008).

[6] 'a sort of low-level background terrorism', according to J.L. Heilbron, Galileo, Oxford University Press (2010).

[7] *Galileo and the Law of Inertia*, by S. Drake, Amer. J. Phys. **32**, 601 (1964).

[8] Moon Man: What Galileo saw, by A. Gopnik (*The New Yorker*, February, 2013).

[9] Footnote 3, p. 216

[10] How Galileo could have derived the special theory of relativity, by A. Sen, *Amer. J. Phys.* **62**, 157 (1994).

[11] Dante's insight into galilean invariance, by L. Ricci, *Nature* **434**, 717 (2005).

[12] At the Venetian Arsenal, where Galileo was a consultant in 1593.

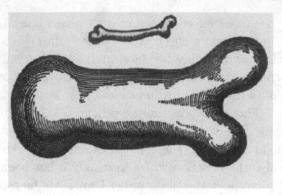

FIGURE 1.2
A sketch depicting the scaling of weight-supporting bones according to
Galileo (reprinted from *Men of Physics: Galileo Galilei, His Life and
His Works*, by R.J. Seeger, Pergamon Press 1966 with permission from
Elsevier).

The dialogue style was very popular in Galileo's time in books explaining scientific matters[13]. In fact, this style of dialogue provided
Galileo an unique medium to improve his propositions, and consider the
objections of his ideas from different perspectives. For example, through
this dialogue, Galileo proposed that Nature is not scale invariant that
led to his discovery of the scaling laws[14]. Galileo pointed out that it is
impossible for giant men ('ten times taller than an ordinary man') to
exist amongst us because that would require the bone materials that
must be stronger and harder than usual (there is even a picture of a
giant bone in the book (Fig 1.2)). The scaled up version of any living or
inanimate objects currently available will simply not be able to survive:
because of 'the mere fact that it is matter that makes the larger machine,
built of the same material and in the same proportion as the smaller,
correspond with exactness to the smaller in every respect except that it
will not be so strong ... Who does not know that a horse falling from a
height of three or four cubits will break his bones, while a dog falling
from the same height or a cat from a height of eight or ten cubits will
suffer no injury? Equally harmless would be the fall of a grasshopper
from a tower or the fall of an ant from the distance of the moon.' When

---

[13] *Galileo at Work: His Scientific Biography*, by S. Drake (Dover Publications, New
York, 1978).
[14] Galileo's discovery of scaling laws, by M.A. Peterson, *Amer. J. Phys.* **70**, 575
(2002).

scaling up, we cannot simply make things bigger without considering the tensile strength of the building materials. A giant of immense height will eventually collapse under its own weight. Scaling is an important part of the field of engineering where a full-scale prototype is often not very practical. The Two New Sciences is an invaluable source of knowledge even for today's technology.

Incidentally, due to the diktat by the Inquisition, publication of this valuable book in Italy was by no means possible. The manuscript was smuggled out of Italy and after the failure of the initial attempts to publish Two New Sciences in France, Germany, and Poland, it was published by Lodewijk Elzevir (the forerunner of the publisher Elsevier of today) in Leiden, Holland, where the writ of the Inquisition was of little consequence. The printed book reached Rome in January 1639 and was 'sold out' (all 50 or so copies) immediately, causing no further harm to the author.

Galileo's scaling law sealed the giant's fate. But what about going in the other direction, viz. scale down to the miniature world? For more than 50 years, miniaturization has driven many of the technological advances that has transformed our daily lives so profoundly. After the invention of the first transistor in 1947, it has shrunk at a pace that is truly incomprehensible. The earliest products that used transistors were portable transistor radios where only a few transistors (< 10) were used (often that number was mentioned on the radio case as a form of advertisement). The transistors in the early days were small drawing-pin-size three-pin devices that were plugged into perforated circuit boards and connected by soldering wires between the transistor and other components. The situation changed dramatically in the 70s when microprocessors, the brain of all electronic devices[15], appeared in the market that contained a few thousand transistors. The number of transistors in today's microprocessors is in tens of billions.

With these rapid technological advances, and the rise of internet, the world has become truly small, where we exchange messages around the world with the press of a button, use the internet for everything including news, music, transportation, shopping, and so many other consumer needs that were unimaginable even a few decades ago. A smart phone that is in almost everyone's hand today, is a combination of PC,

[15] *The Conquest of the Microchip*, by Hans Queisser (Harvard University Press 1988), *The Microprocessor: A Biography* by M.S. Malone (Springer-Verlag, N.Y. 1995).

media player, GPS, a digital camera, a whole host of other features, and of course, the phone, that fits into your pocket. Miniaturization has rapidly improved data storage, with USB flash drives, SD cards, micro-SD cards with the capacity that was unimaginable only a few decades ago, (think e.g., floppy disks)!

Deeply embedded in everything electronic, transistors have permeated our modern life almost as thoroughly as water permeates the sand. Portable, cheap, and reliable devices based on this remarkable object have found their ways in almost every corner of the world, and every facet of our life. The Information Age has been made possible by this nanoscale object that has intimately connected the World and made it a global society[16]. The progress has been so rapid and dramatic that some even think about the possible occurrence of a 'technological singularity', a point in this exponential growth, when technological evolution of mankind will cease and intelligent machines will make discoveries too complex for human endeavor[17]. Perhaps, it is an extreme case, and as improbable as another ominous prediction in modern days that the male Y chromosome is getting progressively smaller and men are heading[18] for extinction! Highly improbable, but they certainly give us pause for thought.

As we will soon find out, scaling down of semiconductor electronics is not only the change in geometry of the devices, but it also involves a radical change in the physical phenomena that alters the properties of the various components of the devices and controls the miniaturization process. It is a bit more complicated than, say, following the example from a classic book: Alice's Adventures in Wonderland[19] by Lewis Carroll. Here the protagonist Alice fell down a rabbit hole only to find a tiny door that leads to a beautiful garden and a golden key on a glass table. Alice shrinks down to ten inches height after drinking a mysterious

---

[16]In that process, it has also taken away our privacy in every imaginable (and unimaginable) way. See e.g., *Blown to Bits: Your Life, Liberty, and Happiness After the Digital Explosion*, by Hal Abelson, Ken Ledeen, and Harry Lewis (Addison-Wesley 2008); *Privacy is power: Why and How You Should Take Back Control of Your Data*, by Carissa Véliz (Bantam Press, UK 2021).

[17]*The Digital Mind: How Science is Redefining Humanity*, by Arlindo Oliveira (MIT Press, 2017).

[18]*Adam's Curse, A Future Without Men*, by Bryan Sykes (W. W. Norton & Co, 2005).

[19]Alice's journey has been called a metaphor for nanotechnology, in *Nanoculture: Implications of the New Technoscience*, ed. N. Katherine Hayles (Bristol: Intellect Books, 2004).

liquid from a bottle labeled 'Drink Me'. Afterwards, to get out of that terrible situation, she eats a bite of a cake that was labeled 'Eat Me' that made her grow to a giant size. Later on, in that story, Alice met a caterpillar sitting on a mushroom, smoking hookah. The caterpillar, before it crawls away into the grass told Alice that eating one side of the mushroom will make her taller while the other side will make her shorter.

Fortunately, we do not need to drink any magic liquid or eat any magic mushroom in order to enter the nanoworld in the next section. The golden key there will be the quantum mechanics.

Apropos the term 'nano' in nanoscience or nanotechnology is derived from the Greek word 'nanos' meaning dwarf.

2

Down to low dimensions

Nanoscience deals with objects on a scale of 1-100 nm. But what is the size of a nanometer? A nanometer is billionth of a meter (1 nm = 0.000000001 m). A distance that is 50,000 times smaller than the thickness of a human hair! Other popular examples include: a strand of human DNA is 2.5 nm in diameter. Ten hydrogen atoms would line up in 1 nm range. At present, the transistors have shrunk to 10-20 nanometers. A bacterium is about 1,000 nanometers. The coronavirus virions are spherical with diameters of approximately 125 nm[1]. These examples perhaps provide a mental picture of how small a nanometer is. At this scale, the matter can be manipulated, atom by atom, to improve its physical or electronic properties. That kind of probe of the nanoworld became a reality with the inventions of scanning tunneling microscope in 1981[2] and the atomic force microscope in 1986[3], which made atomic scale images possible. The invention of the former earned Binning and Rohrer the Nobel Prize in Physics in 1986.

2.1 The essential toolkit: Quantum mechanics

In the nanoscale regime, materials can have dimensions lower than three. There are two-dimensional, one-dimensional, and even zero-dimensional, objects present, either naturally or artificially created. Those are the objects that will be dealt with in the present book. However, the very concept of low-dimensional objects naturally defies our imagination because, while in mathematics, planes, lines, and dots are familiar ideas,

---

[1]Coronaviruses: An overview of their replication and pathogenesis, by A.R. Fehr and S. Perlman, *Methods Mol. Biol.* **1282**, 1 (2015).

[2]Tunneling through a controllable vacuum gap, by G. Binnig, H. Rohrer, Ch. Gerber, and E. Weibel, *Appl. Phys. Lett.* **40**, 178 (1982).

[3]Atomic Force Microscope, by G. Binnig, C.F. Quate, and Ch. Gerber, *Phys. Rev. Lett.* **56**, 930 (1986).

DOI: 10.1201/9781003090908-2

real world is made up of only three-dimensional objects. How thin can the electron plane be, in order to call it two-dimensional? In fact, the Heisenberg uncertainty principle in quantum mechanics[4] would make it difficult to make the electron plane to be arbitrarily thin, because then the uncertainty in momentum in the direction of the thickness would be large and the electron will no longer remain confined[5].

2.2   Two-dimensional electron gas

Interestingly, the low-dimensional structures that are commonly created in order to study physics at the nanoscale are not really less than three dimensions in a strict geometrical sense. The idea here is that once the thickness of the plane where the electrons reside are less than some critical value, quantum mechanics shows that the relevant physical properties become essentially independent of the thickness. This is a consequence of what is known as quantum confinement. A quantum confined structure is one where the electron motion is confined in one or more directions by potential barriers. The consequence of confinement is that the electrons do not have a continuous energy spectrum, but only allowed certain discrete energy levels[6] with large gaps between those levels. Quantum mechanics predicts that, in general, the energy spectrum of electrons in any finite-size systems should be discrete. However, in macroscopic systems, the discreteness of the energy levels cannot be resolved because the energy gaps between two levels is very small. In semiconductor nanostructures, on the other hand, the electrons can be confined to very small spatial regions, resulting in a sizable energy gap. Then the quantization of the energy spectrum plays a very important role. In that case, a 'particle in a box' calculation in quantum mechanics (see Box 2.1)

---

[4] *Quantum Physics in the Nanoworld*, by Hans Lüth (Springer, Heidelberg 2009).

[5] Hesienberg's uncertainty principle, the most fundamental principle of quantum mechanics, states that the uncertainty in position of a particle and the uncertainty in its momentum can never be less than one-half of the reduced Planck constant: $\Delta x \Delta p \geq \hbar/2$. In other words, any effort to reduce the error in our measurement of a particle's momentum will increase the error in our measurement of its position. The two error sizes, when multiplied together, will always exceed a certain value.

[6] *Quantum Semiconductor Structures*, by C. Weisbuch and B. Vinter (Academic Press, 1991); *Electronic States and Optical Transitions in Semiconductor Heterostructures*, by F.T. Vasko and A.V. Kuznetsov (Springer, Heidelberg 1999).

explains how the electrons are forced to remain in the lowest energy levels at certain densities and at very low temperatures. In particular, there are two parameters, the thickness of the electron plane and the average separation of the electrons that characterize the electron confinement in the lowest subband (see Box 2.1)[7]. The electrons, in that situation do not have the required energy to jump across the energy gap to higher energy levels. With only two degrees of free motion being available, the electrons in each quantum state form a two-dimensional (2D) electron system. They are sometimes referred to in the literature as 'dynamically' two-dimensional systems[8]. Recently, a novel class of two-dimensional systems stemming from graphene and other related objects (see Chapter 5) have received a lot of attention, which are just a single layer of atoms where electron motion is strictly two dimensional.

Now that the background principle of what it takes to form the 2D electron system is understood, the challenge is to implement that idea in actual devices. The process by which one makes a 2D electron gas in a real system is quite ingenious. We describe here two different transistor-like structures that are most commonly used in the laboratories: the metal-oxide-semiconductor field-effect transistors (MOSFETs), and semiconductor heterostructures. The two-dimensional electron systems created in those devices were crucial to make the epoch-making discoveries to be described below.

Box 2.1 Quantum confinement:

Let the electron be in a infinitely deep square well as depicted in Fig. 2.1. The electron is free to move in the region $0 < z < a$, and face infinitely high potential barrier to prevent it from straying beyond this region. Due to the constraint of the electron motion, its energy is 'quantized' (found by solving the corresponding Schrödinger equation) and is given as $E_n = p_z^2/2m = \hbar^2\pi^2 n^2/2ma^2$, where p_z is the z-component of the electron momentum, m is the electron mass, and the integer $n = 1, 2, \ldots$ is the 'quantum number' that labels the discrete energy levels that are available to the

---

[7]Physics in less than three dimensions, by L.J. Challis, *Contemp. Phys.* **33**, 111 (1992).

[8]Electronic properties of two-dimensional systems, by T. Ando, A.B. Fowler, and F. Stern, *Rev. Mod. Phys.* **54**, 437 (1982).

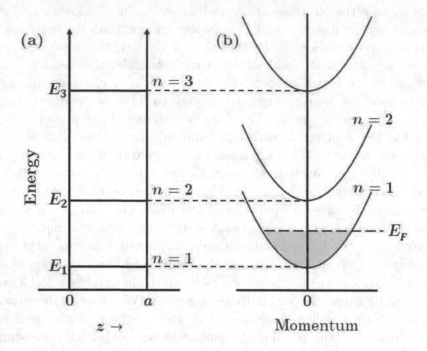

FIGURE 2.1
Energies available to an electron (schematic) (a) in the z-direction, and
(b) the total energy. Electrons fill all the available states below the Fermi
energy and behave dynamically as two-dimensional.

electron. The available energy for the electron is no longer continuous, as expected for electrons that move freely, but only allowed
the discrete set of energies [Fig. 2.1 (a)]. On the other hand, the
motion in the xy-plane is that of a bulk and the electron energy
is $E_{xy} = p_{xy}^2/2m$. The total electron energy of the electron is
then, $E = \hbar^2\pi^2 n^2/2ma^2 + p_{xy}^2/2m$, as shown in Fig. 2.1 (b). Each
parabolic energy curves are called 'subbands'. The 'Fermi energy'
$E_F = (\pi\hbar^2/m)\, n_{2D}$, where n_{2D} is the electron density (the number
of electrons per unit area) and is the highest occupied energy level
at absolute zero temperature, that is, when the electrons occupy
the lowest energy levels permitted by the Pauli exclusion principle.
For a semiconductor, such as GaAs, $E_F \ll E_2$ at absolute zero

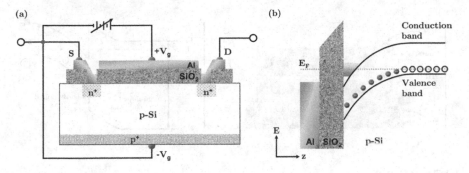

FIGURE 2.2
Schematic view of an Si-MOSFET (left) and the energy level diagram
(right).

temperature. This condition of inequality can be expressed in terms
of the thickness of the electron plane a as $a << d = (\pi n_{2D})^{-1/2}$,
where d is the average separation of the electrons. Interestingly,
even at room temperature, the thermal energy $k_B T \ll E_2 \quad E_F$.
Here, k_B is the Boltzmann constant. Therefore, at low temperatures,
electrons are forced to stay in the lowest level and do not have
sufficient energy to jump to the next higher level. Their physical
properties therefore remain essentially two-dimensional.

Si-MOSFET *structures:* A schematic picture of the device is shown
in Fig. 2.2 (a). It consists of a semiconductor (p-Si) that has a plane
interface with a thin film of insulator (SiO_2), the opposite side of which
has a metal (gate) electrode. Hence the acronym MOS (Metal-Oxide-
Semiconductor). The term 'Field Effect Transistor' (FET) derives from
the effect of the electric field applied to the gate: Application of a voltage
('gate voltage' V_G) between the gate and the Si/SiO_2 interface bends
the electron energy bands. For a strong enough electric field, as the
bottom of the conduction is pushed down below the Fermi energy E_F
(the highest occupied energy level), electrons accumulate in a 2D quasi-
triangular potential well close to the interface [Fig. 2.2 (b)]. Since the
width of the well is small ($\sim$5 nm), the electron motion perpendicular to
the interface is quantized, but parallel to the interface the motion is that
of free electrons, which is exactly the quantum confined state we describe

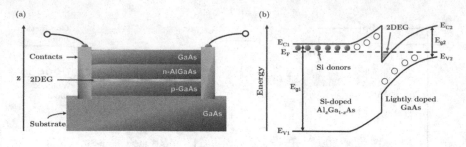

FIGURE 2.3

Schematic view of the heterostructure (left) and the energy diagram at a GaAs-heterostructure interface (right). Note the huge difference between the energy gaps of the two semiconductors: E_{g1} for AlGaAs and E_{g2} for GaAs.

above (Box 2.1) that is essential to create the two-dimensional electrons. This system is called an inversion layer because here the charge carriers are electrons while the semiconductor is p-type. At low temperatures (the thermal energy $k_B T \ll \Delta E$, the energy spacing) the electrons are trapped in the lowest subband and the system is, as expected, purely two-dimensional. This is one of the most important sources for obtaining the two-dimensional electron gas that is the crucial component for the ground-breaking discoveries described in this book.

Some interesting facts about MOSFET: It is the basic unit of the present day ultra-large-scale-integration microelectronics industry. It was the first transistor that could be miniaturized to the extreme and mass-produced that has fuelled the revolution of the electronics industry and thereby the economy, culture, and our daily life. Some estimates suggest that, nowadays, these transistors are produced at the rate of 150 trillions per second. There are about 10^{22} MOSFETS manufactured thus far. More transistors than there are synapses in the brains of all human beings currently alive[9]!

---

[9] *Digital Mind: How Science is Redefining Humanity*, by Arlindo Oliveira (MIT Press, 2017).

Electrons in semiconductor heterostructure interfaces: Superior quality (i.e., containing less impurity) planar electrons are also created in
semiconductor heterostructures at a nearly perfect lattice-matched semi-
conductor/semiconductor interface. One such very widely used system is
the $GaAs/Al_xGa_{1-x}As$ $(0 < x \leq 1)$ heterojunction. The lattice parameters of the two materials are almost the same, but the band gap of the
alloy (E_{g1}) is wider than that of GaAs (E_{g2}) and it increases with the
aluminum concentration x (Fig. 2.3). Electrons in the neighborhood of
the heterojunction move from the doped alloy across the interface to the
low-lying band edge states of the narrow band gap material, viz. GaAs.
The energy bands bend as shown in Fig. 2.3 (b), due to the electric
field from the charge transfer and a quasi-triangular potential well (10
nm), is formed in GaAs that traps the electrons. These structures have
very high electron mobilities, thanks to a technique called 'modulation
doping' pioneered by Horst Störmer and his colleagues[10]. Devices based
on this structure can be used to much higher frequencies than silicon
devices due to the high mobility of electrons in GaAs. These devices are
widely used in cell phones, satellite receivers, etc. For our present purposes, this structure provides the essential source for the high-mobility
two-dimensional electron gas.

2.3 Novel phenomena in flatland: Nobels galore

What heppens if we subject the 2D electron system that we have just
described above to a 'strong' perpendicular magnetic field and extremely
low temperature (typically at liquid Helium temperature T = 4.2 K)?
As it will be shown below, it led to several incredible discoveries and important milestones in the field that has changed the course of research
in condensed matter physics for the past four decades. But first let us
clarify what is 'strong' in this context. We have in mind a magnetic field
of about 20 Tesla or higher. This is to be compared to a typical refrigerator magnet (0.005 Tesla) or the magnet of the MRI in a hospital (1.5-3
Tesla). Schematically, the experimental setup is sketched in Fig. 2.4. The
magnetic field B is perpendicular to the electron plane along which the

---

[10]Electron mobilities in modulation-doped semiconductor heterojunction superlattices, by R. Dingle, H.L. Störmer, A.C. Gossard, and W. Wirgmann, *Appl. Phys.
Lett.* **33**, 665 (1978).

(a)   (b)

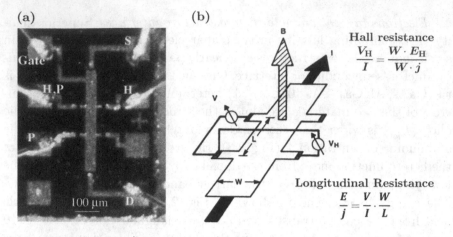

FIGURE 2.4

A typical silicon MOSFET device used for measurements of the Hall effect. For a fixed current between the source (S) and drain (D) contacts, the potential differences between the contacts P-P and H-H are directly proportional to the usual (longitudinal) resistances $R_{xx} = V/I$ and its Hall resistance $R_{xy} = V_H/I$, respectively.

current I flows. The idea is to measure the Hall resistance R_H (resistance across the electron plane) and R_{xx} (resistance parallel to the electron plane) for different magnetic fields and gate voltages (introduced above). On February 5, 1980 at around 2 a.m. during such an experiment at the High Magnetic Field Laboratory in Grenoble, France, Klaus von Klitzing discovered a surprising effect that earned him the Nobel Prize in physics in 1985 (Fig. 2.5)[11].

2.3.1   Quantum Hall effect

In an Si-MOSFET device (Fig. 2.4) subjected to a magnetic field of 19.8 Tesla and liquid helium temperature, von Klitzing noticed that, R_{xx} vanishes at different regions of the gate voltage V_G indicating a current flow without dissipation (Fig. 2.6). The electrons flow without any resistance at those special values of the gate voltage. That is already quite an interesting effect, but more surprisingly, in those regions of the gate voltages, the Hall resistance R_{xy} develops *plateaus* with $R_{xy} = h/ne^2$,

---

[11]The Quantized Hall Effect, by K. von Klitzing, *Rev. Mod. Phys.* **58**, 519 (1986).

FIGURE 2.5
Discovery of the quantum Hall effect by Klaus von Klitzing (photo courtesy of Dr. von Klitzing).

where n is an integer, h is the Planck's constant and e is the elementary charge. The Hall resistance is therefore quantized at those values of the gate voltage where the parallel resistance goes to zero. What is even more astonishing, the quantization condition of the plateaus that just depend on two fundamental constants, are found to be obeyed with *extreme accuracy*, independent of the material, geometry of the sample, and microscopic details of the semiconductor[12]. In the absence of the magnetic field, i.e., $B = 0$, the effect is totally absent. This remarkable discovery opened the floodgate for more novel discoveries and forty years of unceasing activities by numerous researchers that is yet to subside[13].

How does one explain such an extraordinarily accurate phenomenon in a system that is normally considered to be dirty (the unabashed use of the term 'doping' in these devices should convince everyone about it). Again, the quantum mechanics comes to the fore to explain the

---

[12]New method for high-accuracy determination of the fine-structure constant based on quantized Hall resistance, by K.v. Klitzing, G. Dorda, and M. Pepper, *Phys. Rev. Lett.* **45**, 494 (1980).

[13]40 years of the quantum Hall effect, by K. von Klitzing, T. Chakraborty, P. Kim, et al., *Nature Rev. Phys.* **2**, 397 (2020).

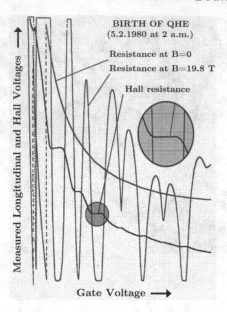

FIGURE 2.6

The first experiment showing the quantized-Hall effect. Hall resistance and longitudinal resistance (at B = 0 and B = 19.8 Tesla) of a silicon MOSFET at liquid helium temperature (4.2 Kelvin) versus the gate voltage. The Hall resistance is seen to be quantized, while at those values of the gate voltage the resistance parallel to the plane vanishes. The enlarged part depicts the Hall plateau at filling factor $\nu = 4$.

effect. In the quantum mechanical picture, free two-dimensional electrons subjected to a strong perpendicular magnetic field B are allowed only discrete energy levels, the so-called Landau levels[14]. These energies are expressed as $E_n = \left(n + \frac{1}{2}\right) \hbar \omega_c$, $n = 0, 1, 2, ...$, where $\omega_c = eB/m$, m is the electron mass, $h = 2\pi\hbar$ is the Planck constant. The energy levels are separated by what is known as the Cyclotron energy $\hbar \omega_c$, which increases with increasing magnetic field (Fig. 2.7). Each Landau level (LL) can accomodate large number of electrons (a property known as Landau level degeneracy, i.e., having the same energy). The number of available states for electrons in each LL is given by $n_L = B/\Phi_0$, where

---

[14] *Bemerkung zur Quantelung des harmonischen Oszillators im Magnetfeld*, by V. Fock, Zeit. f. Phys. **47**, 446 (1928); Diamagnetismus der Metalle, by L. Landau, *Zeit. f. Phys.* **64**, 629 (1930); The Diamagnetism of the free electron, by C.G. Darwin, *Math. Proc. Cambridge Phil. Soc.* **27**, 86 (1931).

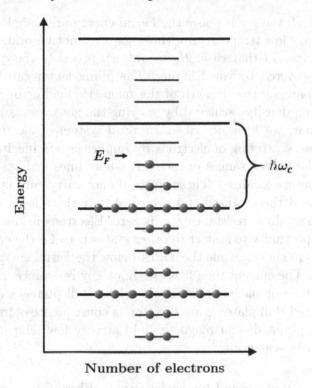

Number of electrons

FIGURE 2.7
Schematic filling of Landau levels (long lines) by electrons upto the Fermi energy E_F. The short lines are the localized states due to the impurities. The empty states are high up in energy and cannot be reached by the electrons at very low temperatures. In the absence of any scattering, electrons flow without any resistance, while the number of filled Landau levels explain the quantization of the Hall plateaus.

$\Phi_0 = h/e$ is called the magnetic flux quantum. The number of filled LLs, i.e., the 'filling factor' is $\nu = n_{2D}/n_L = n_{2D}h/eB$, where n_{2D} is the areal density of electrons. For a homogeneous 2D electron system, the Hall resistivity is $R_H = B/en_{2D} = h/(\nu e^2)$.

At extremely low temperatures, the electrons populate all the states below the Fermi energy (which depends on the areal electron density), while all the states above the Fermi energy are completely empty (Fig. 2.7). Now, if the Fermi energy lies in the gap between two LLs (i.e., $\nu = n$), then all the n LLs below the Fermi energy are totally filled, while all the levels above the Fermi energy are completely empty. At sufficiently low temperatures, those gaps cannot be bridged by thermal scattering. In that case, $R_{\mathrm{H}} = h/ne^2$, precisely the quantization relation discovered by von Klitzing. The filling factor can be changed either by changing the strength of the magnetic field or by varying the areal electron density (achieved by varying the gate voltage).

Until now, we have described an *ideal* system of electrons. In actual samples, scattering of electrons by impurities are inevitable, which broadens the energy ranges in each LL (short lines in Fig. 2.7), where the electrons are localized. Those states do not carry current[15]. The energy gaps and the localized states ensure that there is no dissipation, i.e., the longitudinal resistance R_{xx} is zero. Electrons in the filled levels have no opportunity to scatter to other states because the empty states are high up in energy, while the states below the Fermi energy are completely full. The current then flows without any resistance in the region of gate voltage or magnetic field, where the Hall plateaus are formed. The quantized Hall plateaus are therefore a consequence of free electrons in a strong perpendicular magnetic field at very low temperatures and weak impurity scattering[16].

2.3.2   A new standard for resistance calibration

Metrology, the science of measurement, has been around since ancient times in order to assist in trades and exchanges. It was developed as far back as 6,000 B.C. with the development of agricultural settlements. The modern day system of measuring units that are internationally agreed upon, dates back to around late nineteenth century[17]. The integer quantum Hall effect (IQHE), as the discovery is popularly named, found an important application in metrology, where the effect is used to represent a resistance standard. The epoch making

---

[15]Quantized Hall resistance and the measurement of the fine-structure constant, by R.E. Prange, *Phys. Rev. B* **23**, 4802 (1981).

[16]Here and in what follows, we have adapted the 'quantum Hall lite' version in order to appeal to the non-experts. For a deeper understanding of the exact quantization, the reader is encouraged to consult, e.g., Quantized Hall conductivity in two dimensions, by R.B. Laughlin, Phys. Rev. B **23**, 5632 (1981); Quantized Hall conductance, current-carrying edge states, and the existence of extended states in a two-dimensional disordered potential, by B.I. Halperin, *Phys. Rev. B* **25**, 2185 (1982).

[17]A brief history of metrology: past, present, and future, by J.-P. Fanton, *Int. J. Metrol. Qual. Eng.* **10**, 5 (2019).

discovery was the 'exact quantization' of Hall resistance to a fundamental value of $h/e^2 = 25812.807...\,\Omega$ that is incredibly robust. The
experimental accuracy so far achieved is nearly one part in a billion.
This has facilitated the definition of a new practical standard for electrical resistance based on the resistance quantum given by the *von Klitzing
constant* $R_{\mathrm{K}} = 25812.807449 \pm 0.000086\,\Omega$. Since 1990, a fixed value of
$R_{\mathrm{K}-90} = 25812.807\,\Omega$ has been adopted internationally as a standard
for resistance calibration[18].

Table I: Summary of high precision data for the quantized Hall
resistance up to 1988 which led to the fixed value of 25 812.807 Ω
recommended as a reference standard for all resistance calibrations
after 1.1.1990.

(Hall-) Resistance	$R_{\mathrm{H}}(\Omega)$
PRL **45**, 494 (1980)	25 812.68 (8)
BIPM (France)	25 812.809 (3)
PTB (Germany)	25 812.802 (3)
ETL (Japan)	25 812.804 (8)
VSL (The Netherlands)	25 812.802 (5)
EAM (Switzerland)	25 812.809 (4)
NBS (USA)	25 812.810 (2)
NPL (UK)	25 812.811 (2)
1.1. 1990	**25 812.80700**

Table I provides a summary of high precision data for the quantized Hall resistance until 1988 that led to the fixed value of $R_{\mathrm{K}-90}$
recommended as a reference standard for all resistance calibrations after[19] January 1, 1990. The von Klitzing constant has provided perhaps the most accurate measurement of the fine-structure constant
$\alpha = (e^2/h)\mu_0 c/2 \approx 1/137$, introduced into physics by A. Sommer-
feld in 1916. It is a dimensionless constant which means it is simply
a pure number and contains a group of constants: elementary charge e,
Planck's constant h, speed of light c and the permeability of vacuum

---

[18]The quantum Hall effect as an electrical resistance standard, by B. Jeckelmann
and B. Jeanneret, *Séminaire Poincaré* **2**, 39 (2004).

[19]Taking stock of the quantum Hall effects: Thirty years on, by T. Chakraborty
and K. von Klitzing, *Physics in Canada* **67**, 161 (2011).

μ_0, and is a fundamental constant that characterizes the strength of electromagnetic interaction between elementary charged particles. It is 'one of the fundamental constants of Nature characterizing a whole range of physics including elementary particles, atomic, mesoscopic and macroscopic systems'[20].

In 2019, the SI underwent a major revision on the World Metrology day (May 20) that was implemented by the international measurement community. This global change involved moving away from material artefacts and instead employing a set of seven fundamental constants of Nature, from which all units of measure can be derived. The seven base units associated with the constants (Hyperfine transition frequency of Cs, Speed of light in vacuum, Planck constant, Elementary charge, Boltzmann constant, Avogadro constant, and Luminous efficacy) are now directly linked to seven fixed values, and four of which (the Planck constant (h), the elementary charge (e), the Boltzmann constant (k_B) and the Avogadro constant (N_A)) have been modified to represent an exact value[21].

The changes to constants h and e have direct impacts on how the ohm, volt, and ampere are defined. This has led to the von Klitzing constant changing from its conventional value, set in 1990 ($R_{K-90} = 25812.807\,\Omega$) to the fixed numerical values of the defining constants[22] h/e^2 ($R_K = 25812.8074593045\,\Omega$).

2.3.3   Quantum Hall effect – now with fractions

After the spectacular discovery of the integer QHE and its subsequent explanation, one would have thought that the chapter on quantum Hall effect is perhaps closed. Surprisingly, however, the situation was exactly the opposite and we were soon overwhelmed with many more novel and profound discoveries in this field and many more Nobel prizes followed. After just two years of the IQHE discovery, Horst Störmer and his collaborators discovered what is now known as the fractional

---

[20]The fine structure constant, by T. Kinoshita, *Rep. Prog. Phys.* **59**, 1459 (1996).

[21] *The New International System of Units (SI): Quantum Metrology and Quantum Standards*, by E.O. Göbel and U. Siegner, (Wiley-VCH Verlag, Weinheim, 2019); The revision of the SI – the result of three decades of progress in metrology, by M. Stock, et al., *Metrologia* **56**, 022001 (2019).

[22]Essay: Quantum Hall effect and the new International system of units, by K. von Klitzing, *Phys. Rev. Lett.* **122**, 200001 (2019).

quantum Hall effect (FQHE)[23] in 1982. The experimental results look uncannily similar to that of the IQHE, but as it is often said, looks can be very deceiving. In a high-mobility 2D electron gas in GaAs/AlGaAs heterostructure, grown at the famous Bell Laboratories in New Jersey, at much lower temperatures and much stronger magnetic fields, they discovered Hall plateaus at $h/(1/3)e^2$ and $h/(2/3)e^2$ (in contrast to the integer values discovered by von Klitzing), accompanied by the vanishing longitudinal resistivity thus revealing again a dissipationless current flow at those values of the magnetic field (Fig. 2.8). The fractionally quantized Hall effect is totally unexpected and not so easy to comprehend.

In the case of the integer quantum Hall effect, we explained the vanishing of R_{xx} due to the presence of the Landau gap and the presence of Hall plateaus due to the degeneracy of the filled Landau levels. Clearly, that is prima facie impossible in the present case because there are no Landau energy gaps below the lowest Landau level, and one must explain the presence of a gap that is responsible for FQHE due to some other sources.

As the plateaus and minima appear at fractional filling factors where no structure was expected in a free-electron picture that was used to explain the integer effect, it was clear that one must look for a situation where many electrons contribute collectively, i.e., the electron-electron interaction must play a crucial role in generating the required gap. That is a rather challenging problem as there are at least 10^{15} electrons per square meter present in those devices and developing a theory to describe the properties of a system of that many number of electrons is a formidable assignment.

Among various attempts to obtain such a many-body state, the most successful theory was proposed by Robert Laughlin in 1983. Just as the experimental discovery was a major surprise, Laughlin's theory was equally astonishing and unique. Many thousands of papers have been published following the brilliant ideas of Laughlin. The impact of the experimental and theoretical works on the field of novel quantum fluid was recognized by the Nobel Prize for Störmer, Laughlin, and Tsui in 1998. It is certainly not possible to cover in this book most of those theoretical and experimental works reported until this day. Instead, we shall

---

[23]Two-dimensional magnetotransport in the extreme quantum limit, by D.C. Tsui, H.L. Störmer, and A.C. Gossard, *Phys. Rev. Lett.* **48**, 1559 (1982).

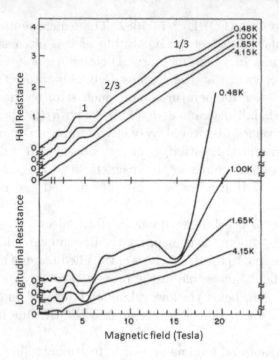

FIGURE 2.8
Discovery of the fractionally quantized-Hall effect at $\nu = \frac{1}{3}$, i.e., the Hall resistance shows a plateau at $3h/e^2$ (and a weak $\nu = \frac{2}{3}$).

focus on the works directly related to Laughlin's theory and developed in the early days of the discovery.

2.4   Laughlin's eponymous wave function

Investigation of the properties of systems containing many electrons is one of the most intractable problems of quantum physics, especially when the mutual interactions among the electrons are also included. There are several theories developed in the past[24] to evaluate the

---

[24]Correlated electrons in quantum matter, by Peter Fulde, *World Scientific*, Singapore, 2012; *Many-Electron Theory*, by S. Raimes, North-Holland, Amsterdam 1972; *The Many-Body Problem in Quantum Mechanics*, by N.H. March, W.H. Young, and S. Sampanthar, Cambridge University Press, 1967; Coulomb Liquids, by N.H. March and M.P. Tosi, Academic Press, 1984.

properties of interacting electrons, viz., the ground state (lowest) energy, the pair correlation functions[25], etc. of the interacting electron system in the 'thermodynamic limit'[26]. In this regard, the Jastrow variational approach has been particularly useful[27]. However, Laughlin's theory is quite unique in that respect and it goes far beyond the conventional theories we are familiar with until now.

The best source for understanding Laughlin's approach is, of course, his own[28] articles. Interestingly, the most comprehensive account of Laughln's theory (not least the mathematical details about its relation to the classical one-component plasma, and the many-body hypernetted-chain formalism) can still be found in our 1988 book[29]. In his theory, Laughlin demonstrated that the ground state at the filling factor $\nu = \frac{1}{3}$, where the effect was observed in the experiment, is that of a uniform density liquid with several atypical properties. He also explained the mechanism behind the exceptional stability of the $\frac{1}{3}$ state, a reason why it is detected in the experiment as the strongest FQHE effect.

2.4.1 The (lowest energy) ground state

The electrons are confined in the xy-plane which is taken to be a complex plane with $z = x - iy$ being the electron position. The magnetic field

---

[25]The pair correlation function or radial distribution function $g(r)$ describes quantitatively the internal structure of fluids. It is directly related to the structure factor that can be experimentally determined. See, for example, Electron correlations in inversion layers, by M. Jonson, *J. Phys. C: Solid State Phys.*, **9**, 3055 (1976); Pair correlation function for a two-dimensional electron gas, by M.L. Glasser, *J. Phys. C: Solid State Phys.* **10**, L121 (1977).

[26]In this limit the number of electrons $N \to \infty$, the area $A \to \infty$, but the density $N/A = $ finite.

[27]A new approach to the theory of classical fluids I, by T. Morita and K. Hiroike, *Prog. Theor. Phys.* **23**, 1003 (1960); A new approach to the theory of classical fluids II, by K. Hiroike, *Prog. Theor. Phys.* **24**, 317 (1960); Variational study of the ground state of a Bose-Einstein fluid, by K. Hiroike, *Prog. Theor. Phys.* **27**, 342 (1962); Hypernetted-chain Euler-Lagrange equations and the electron fluid, by John G. Zabolitzky, *Phys. Rev. B* **22**, 2353 (1980).

[28]Anomalous quantum Hall effect: An incompressible quantum fluid with fractionally charged excitations, by R.B. Laughlin, *Phys. Rev. Lett.* **50**, 1395 (1983); Primitive and composite ground states in the fractional quantum Hall effect, *Surf. Sci.* **142**, 163 (1984).

[29]*The Fractional Quantum Hall Effect*, by T. Chakraborty and P. Pietiläinen, (Springer 1988); *The Quantum Hall Effects*, 2nd edition, Springer, 1995. Henceforth we refer to it as the **'QHE book'**.

is strong such that the energy between two Landau levels, i.e., $\hbar\omega_c \gg e^2/\epsilon\ell_0$, the right-hand term being the Coulomb energy unit, where ϵ is a material constant. In this situation, the electrons are in the lowest spin state of the lowest Landau level. Here, $\ell_0^2 = (\hbar/eB)$ is the magnetic length. Guided by his earlier work on two- and three-electrons in the lowest Landau level[30] and in a circularly symmetric system, Laughlin proposed the celebrated ground state (the lowest energy state) many-body wave function that describes the novel quantum state

$$\psi_q\left(z_1, z_2, \cdots, z_N\right) = \prod_{i<j}^{N} \left(z_i - z_j\right)^q \prod_{j=1}^{N} e^{-|z_j|^2/4\ell_0^2}.$$

Electrons are fermions (see Sec. 2.6). In order that the wave function satisfies the Fermi statistics, i.e., the wave function must change sign when two particle positions are exchanged, q has to be an *odd* integer. This parameter q uniquely determines the filling factor $\nu = 1/q$. For $q = 3$ this is the wave function for the filling factor $\nu = 1/3$ state (see Box 2.2) that was observed in the experiment.

Box 2.2 The filling factor:

In the polynomial part of Laughlin's wave function, $\prod_{i<j}^{N}\left(z_i - z_j\right)^q$, an electron coordinate z_i has the maximum power of $(N-1)q \simeq Nq$ for large N. In a circularly symmetric system, that power of z_i also corresponds to the maximum angular momentum quantum number of the electron, and is written as $A/2\pi\ell_0^2$, where A is the area of the circular orbit. This leads to $q = A/2\pi\ell_0^2 N = 1/\nu$ (see Footnote 29, the 'QHE book'). The parameter q is thus fixed by the filling factor and unlike in conventional Jastrow theory (see footnote 27), we do not have any variational parameter to optimize the energy.

For $q = 1$ we have a filled Landau level (IQHE) and the polynomial $\prod_{j<k}\left(z_j - z_k\right)$ is the Vandermonde determinant of order N (see Box 2.4).

---

[30]Quantized motion of three two-dimensional electrons in a strong magnetic field, by R.B. Laughlin, *Phys. Rev. B* **27**, 3383 (1983).

Although it is a very appealing choice as a wave function for $\nu = 1/3$, any practical use of this function was not immediately apparent. This is because finding a suitable wave function of the new state is one thing, but one has also to demonstrate that this function indeed has the lowest energy which is a quite crucial necessity. The difficulty arises due to the explicit dependence of the wave function on the electron positions in the second term on the right-hand side (the Gaussian term) of ψ_q. Any calculation of the equilibrium properties with this wave function would have been almost impossible because that would require evaluation of infinite dimensional integrals. The steps that were needed to circumvent this problem were entirely due to Laughlin's brilliant idea of mapping the system described by this wave function to a two-dimensional, charge-neutral classical plasma (independent of the magnetic field). This mapping between two very different systems not only provided a means to obtain the so called pair-correlation function and with that the ground state energy, it also offered a robust picture of the ground state, as explained below. It provided the explanations for (i) uniform density of the liquid state, (ii) the nature of incompressibility of the liquid, (iii) the 'fractional electron charge' of the excitations, and (iv) calculation of the excitation energies above the ground state. All those properties were totally unique in a many-electron system.

As the one-component plasma has been investigated by several groups very successfully for many years, this mapping gave Laughlin unique opportunities to pick from the relevant results available in that well-established field. The classical plasma analogy also helped in understanding the low-energy excitations in this system in the vicinity of the stable filling factors $\nu = \frac{1}{q}$ as discussed below.

Box 2.3 One-component plasma (OCP):

It is a system of identical electrically charged point particles of charge e and interact with each other exclusively through the Coulomb potential. In order to maintain the overall charge neutrality of the system, the particles are immersed in a rigid uniform background of opposite charge.[31] The quantum counterpart of OCP is the well-known 'jellium' model of the electron fluid.[32] An equilibrium state of the OCP is fully characterized by a single dimensionless parameter $\Gamma = e^2/k_{\mathrm{B}}T$, the 'coupling parameter'. The OCP freezing transition is at a coupling of $\Gamma \simeq 140$. Various many-body techniques have been developed over the years to investigate the equilibrium properties, e.g., the pair-correlation functions $g(r)$, the ground state energies E_0 for this system.[33]

Following the steps taken by Laughlin we write, $\left|\psi_q\right|^2 = e^{-\mathcal{H}_q}$, which gives $\mathcal{H}_q = -2q\sum_{i<j}\ln\left|z_i - z_j\right| + \sum_j \left|z_j\right|^2 /2\ell_0^2$. We immediately recognize $\mathcal{H}_q$ as the Hamiltonian of the OCP with $\Gamma = 2q$ and the charge-neutrality condition of the plasma determines the filling factor to be $\nu = \frac{1}{q}$. The offending Gaussian term in Laughlin's wave function is now recognized as the particle-background interaction term, and can easily be dealt with within the standard many-body scheme. All the pieces of the puzzle now fit nicely as we make use of the pair-distribution function of the OCP at $\Gamma = 2q$, $q > 1$ being an odd integer ($q = 3, 5, \ldots$ etc.), which was already established to correspond to a *uniform density liquid* (OCP freezes at $q \simeq 70$). The electrons however live in a three-dimensional world and the energy of the $\frac{1}{3}$ state needs to be evaluated from the Coulomb interaction. The ground state energy (per particle) is obtained from the expression, $E_0 = \frac{1}{\sqrt{2q}}\frac{e^2}{\epsilon\ell_0}\int_0^\infty [g(x) - 1]\,dx$ with the $g(x)$ determined from the OCP. The second term of the integrand is due to the neutralizing background. The energy was evaluated to be $E_0\left(\nu = \frac{1}{3}\right) = -0.4056\,e^2/\epsilon\ell_0$, which is extremely close to the energy value reported by various numerical schemes (see the QHE book for details). The $q = 1$ case is described in Box. 2.4, and as explained already, is an alternate description of the IQHE, i.e., for a fully filled Landau level.

---

[31] A Monte Carlo study of the classical two-dimensional one-component plasma, by J.M. Caillol, D. Levesque, J.J. Weis, and J.P. Hansen, *J. Stat. Phys.* **28**, 325 (1982).

[32] *Quantum Theory of the Electron Liquid*, by G.F. Giuliani and G. Vignale, Cambridge University Press, 2005.

[33] Integral equation solutions for the classical electron gas, by J.F. Springer, M.A. Pokrant, and F.A. Stevens, Jr., *J. Chem. Phys.* **58**, 4863 (1973); The two-dimensional one-component plasma, by J.P. Hansen and D. Levesque, *J. Phys. C: Solid State Phys.* **14**, L603 (1981).

Laughlin explained the exceptional stability of the filling factor $\nu = \frac{1}{3}$ by showing that the quantum state described by his wave function is in fact, *incompressible*, as explained in the next section below.

Box 2.4 Bite-sized: Stories of $q = 1$

Soon after the integer QHE was discovered by von Klitzing, three physicists from the Landau Institute near Moscow, Yu. A. Bychkov, S.V. Iordanskii, and G.M. Éliashberg reported on February 5, 1981, their work on 2D electrons in a strong perpendicular magnetic field.[34] They followed the path that Laughlin took two years later, of studying two, and three electrons and then reached the final result for $N \to \infty$ number of electrons. Their proposed wave function was: $\psi(\{z_i\}) = \prod_{i<j}^{N} (z_i - z_j)\, \mathrm{e}^{-\sum_{i=1}^{N}|z_i|^2/4\ell_0^2}$, where the first term on the right hand side is the Vandermonde determinant[35] of order N:

$$\prod_{i<k}^{N} (z_i - z_k) = \begin{vmatrix} 1 & 1 & \cdots & 1 \\ z_1 & z_2 & \cdots & z_N \\ z_1^2 & z_2^2 & \cdots & z_N^2 \\ \vdots & \vdots & \ddots & \vdots \\ z_1^{N-1} & z_2^{N-1} & \cdots & z_N^{N-1} \end{vmatrix} .$$

Here z_i is the complex coordinate. The particle density in this state is $1/2\pi\ell_0^2$, i.e., it corresponds to a fully occupied Landau level in the limit of large N. Interestingly, this is a natural precursor to the Laughlin state. Without doubt these authors were very close to achieving Laughlin's wave function (i.e., $q = 3, 5, ...$) but the FQHE was yet to be discovered a year later. Talk about being ahead of their time!

One of the most interesting findings of the one-component classical plasma is that the thermodynamics and the correlation functions can be evaluated *exactly* for one special value[36] of the coupling constant, $\Gamma = 2$. The pair-correlation function in this case can be written

---

[34] *JETP Lett.* **33**, 143 (1981).

[35] A homogeneous polynomial of degree $N(N-1)/2$. It is antisymmetric with respect to the interchange of any pair of variables (z_i, z_j).

[36] On the classical two-dimensional one-component Coulomb plasma, by A. Alastuey and B. Jancovici, *Le Journal de Physique* **42**, 1 (1981); Exact results for the two-dimensional one-component plasma, by B. Jancovici, *Phys. Rev. Lett.* **46**, 386 (1981).

in a closed form, $g(x) = 1 - e^{-x^2}$. A feat that is hard to achieve in most many-body systems. The plasma experts had no explanation for these unique results.

Laughlin explained these remarkable results by pointing out that $\Gamma = 2$ actually corresponds to $q = 1$ in his theory (see Box 2.3), which corresponds to a filled Landau level as described above, and the total energy equals the Hartree-Fock energy, $E = -\sqrt{\pi/8}$ (in units of $e^2/\epsilon\ell_0$). This is the underlying reason for the existence of exact solution at this value of Γ. The second term of $g(x)$ is just the *exchange hole* in the Hartree-Fock theory with the charge of an electron.[37]

Therefore, in addition to proposing the most successful theory of the FQHE, Laughlin also made a significant contribution in the field of classical plasma.

2.5   Incompressible liquid and the charged excitations

The quantum state at $\nu = \frac{1}{q}$ with q being an odd integer (as discovered in the experiment) is exceptionally stable because at that filling factor, there is a positive discontinuity in the chemical potential (energy required to add or subtract a charge to the system), i.e., $\mu_+ - \mu_- > 0$. As a consequence, the electron state is said to be incompressible (see Box 2.5 for details). This also means that there is a downward cusp in energy versus the filling factor which indicates the presence of an energy gap, that is necessary in order to explain the observation of the FQHE. The cusp in energy at integer ν is a consequence of Landau level quantization (Fig. 2.7). For the fractional filling factor the situation is different. Laughlin showed that the elementary charged excitations in a stable $\nu = \frac{1}{q}$ state are quasiholes and quasiparticles with electric charge $\pm\frac{e}{q}$! This is the first example of (quasi-)particles in the field of condensed matter that have *fractional* electron charge and carry a current (see Box 2.6).

---

[37] *Many-Particle Physics*, by G.D. Mahan, Kluwer Academic/Plenum Publishers, N.Y. 2000, 3rd Edition, p. 310.

Box 2.5 Compressibility and the chemical potential

The compressibility in a 2D electron system is defined as $\kappa \equiv -\frac{1}{A}\frac{\partial A}{\partial P}$ where A is the area of the system and P is the pressure. In terms of the chemical potential, $\mu = \partial E/\partial N$, N being the total number of particles, the compressibility is $1/\kappa = n^2 \partial\mu/\partial n = (\nu^2/\pi\ell_0^2)(\partial\mu/\partial\nu)$, where $n = N/A$ is the particle density and $\nu = 2\pi\ell_0^2 n$ is the filling factor. Therefore, $\kappa = 0$ when the chemical potential is discontinuous.

In the *incompressible* liquid, even an infinitesimal change in electron density (or, equivalently the filling factor) costs a finite amount of energy. That explains the stability of the $\nu = \frac{1}{q}$ state. The incompressibility of the 2D electron gas at $\nu = \frac{1}{q}$ means that a non-zero minimal energy is required to move away from that filling factor. In Laughlin's picture, the elementary charged excitations in a stable $\nu = \frac{1}{q}$ state would be quasiparticles and quasiholes with electric charge $\pm e/q$.

Evaluation of the quasiparticle energy is quite involved and highly non trivial[38]. Those calculations indicated that the energy to create the quasihole or the quasiparticle from the Laughlin ground state is certainly non-negligible and provides a strong reasoning for the appearance of the coveted energy gap in the FQHE. This explains the observation of the $\nu = \frac{1}{3}$ quantized Hall effect in the experiment.

Box 2.6 Curioser and curioser: Particles with fraction of an electron charge

---

[38] See Footnotes 28, 29. See also, Elementary excitations in the fractional quantum Hall effect, by T. Chakraborty, *Phys. Rev. B* **31**, 4026 (1985); Monte Carlo evaluation of trial wave functions for the fractional quantized Hall effect: Disk geometry, by R. Morf and B.I. Halperin, *Phys. Rev. B* **33**, 2221 (1986).

In order to introduce a single quasihole to the ground state at the filling factor $\nu = \frac{1}{q}$ (q being an odd integer), that corresponds to the lowest energy charged excitations from the ground state, Laughlin proposed the wave function: $\psi_q^- = \exp\left(-\frac{1}{4}\sum_l |z_l|^2\right) \prod_j (z_j - z_0) \prod_{j<k} (z_j - z_k)^q$, where $z_0 = x_0 - iy_0$ is the quasihole position. It is the Laughlin's original wave function ψ_q multiplied by a factor $\prod_j (z_j - z_0)$, where at z_0 an additional flux quantum has been added to change the filling factor slightly from $\nu = \frac{1}{q}$. The significance of this choice of the proposed wave function can again be found from the classical plasma analogy. Following similar steps that were used in the case of the ground state wave function (see Box 2.3), the plasma Hamiltonian in this case can be found to be that of a classical one-component plasma with an extra phantom point charge at z_0 whose strength is *less* by a factor $1/q$. Since the plasma would like to be charge neutral, it will neutralize the phantom by creating a *deficit* of $1/q$ charge near z_0. Elsewhere in the interior of the plasma the charge density remains unchanged. Crucially, it should be noted that the Gaussian factor in the above quasihole wave function is not affected by the extra function involving the quasihole position and it is responsible for the charge neutrality of the original plasma (i.e., without the additional phantom charge). Hence for the overall charge neutrality of the entire system, the plasma itself creates a deficit of charge near the phantom charge. The actual three-dimensional electric charge is carried by the electrons, and the uniform positive background charge keeps the system charge neutral. Therefore, the real electron system will have a net charge $-e/q$ accumulated in the vicinity of z_0 and a quasihole excitation is thereby created. A somewhat similar situation (but a bit more complex) can be thought of for the fractionally charged quasielectrons.

As an interesting aside: Laughlin's choice of the wave function at $\nu = \frac{1}{q}$ with q being an odd integer cannot be directly applied to the electron system when q is an *even* integer, because in that case the corresponding wave function would describe a system of particles obeying Bose statistics (see on the following page). Laughlin explained that this is the reason why we do not observe FQHE at the filling factor $\nu = \frac{1}{2}$ in the experiment (Fig. 2.8), since electrons are not bosons.

2.6 The unusual statistics

In the quantum world, all particles belong to two mutually exclusive families: bosons (named after Satyen Bose from Calcutta, a quintessential Bengali[39]) and fermions (named after Enrico Fermi, the Italian-American). There are no other types of particles, fundamental or composite, in the entire Universe. They are classified as such depending on whether they obey Bose statistics or Fermi statistics. Bosons (photons, ^{4}He atom, mesons, etc.) have integer spin in units of $\hbar$, and wave functions for many bosons are symmetric under interchange of any pair. Fermions (electrons, protons, neutrons, etc.), on the other hand, have spin equal to half an odd integer and their wave functions are antisymmetric under pairwise interchange. In fact, the statistics determine the structure of the many-body wave functions, which in turn determines the unique behavior of these two groups of particles. Bosons are gregarious and their motto is 'the more the merrier'! Identical bosons like to occupy the lowest energy state with as many members as possible. The manifestation of this important physical consequence is, e.g., the super-fluid nature of ^{4}He even at absolute zero temperature[40]. Fermions, on the other hand, are extremely individualistic and do not share a quantum state with any other identical fermion. They obey the Pauli exclusion principle which naturally follows from the antisymmetry of the fermion wave functions.

The essential reason for wave functions of identical particles to be even or odd under exchange of pair can be explained as follows: Suppose the two identical particles are described by the wave function $\psi(\mathbf{r}_1, \mathbf{r}_2)$. Since the particles are identical when they are exchanged then $|\psi(\mathbf{r}_1, \mathbf{r}_2)|^2 = |\psi(\mathbf{r}_2, \mathbf{r}_1)|^2$, i.e., the probabilities of finding the particles at a given point must be the same. Therefore, upon exchange the

---

[39]Satyendranath Bose: Co-founder of quantum statistics, by W.A. Blanpied, *Amer. J. Phys.* **40**, 1212 (1972).

[40]*Superfluidity*, by E.V. Thuneberg, Encyclopedia of Condensed Matter Physics, (Elsevier, 2005); https://www.scientificamerican.com/article/superfluid-can-climb-walls/

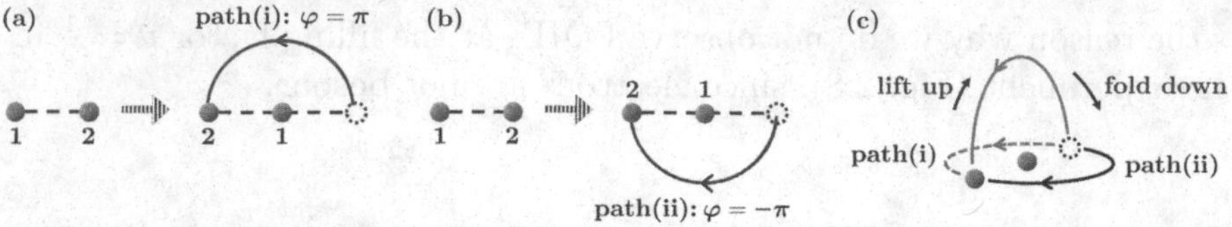

(a)   path(i): $\varphi = \pi$ (b)   (c)

FIGURE 2.9
(a) Counterclockwise exchange of particle positions. (b) Clockwise exchange of particle positions. (c) Folding the path in the third dimension.

wave functions differ by at most a phase, $\psi(\mathbf{r}_1, \mathbf{r}_2) = e^{i\pi\alpha}\psi(\mathbf{r}_2, \mathbf{r}_1)$. The phase acquired by the wave function depends on the parameter α which is usually called *statistics*. If we again exchange the particles, we should be back to where we began, i.e., we make a complete rotation. This means that, $\psi(\mathbf{r}_1, \mathbf{r}_2) = e^{i2\pi\alpha}\psi(\mathbf{r}_1, \mathbf{r}_2)$, and hence $e^{i2\pi\alpha} = 1$.

Let us suppose that the exchange takes place as shown[41] in Fig. 2.9: (i) Move particle 2 around particle 1 by $\varphi = \pi$, and then make a translation of the center of mass to arrive at the initial configuration [Fig. 2.9 (a)].

Here the wave function acquires a phase $e^{i\pi\alpha}$. (ii) Move particle 2 around particle by $\varphi = -\pi$ and then a translation of the center of mass to the original configuration [Fig. 2.9 (b)]. Here the wave function acquires a phase $e^{-i\pi\alpha}$.

In three dimensions, there is no intrinsic difference between the above two cases. One possible way to go from path (i) to path (ii) would be to lift the path (i) in the third dimension and fold it back onto the plane. Then superpose it on to the path (ii) [Fig. 2.9 (c)]. Therefore, the wave function should have the same form, $e^{i\pi\alpha} = e^{-i\pi\alpha}$. This is true only for $\alpha = 0, 1$. Clearly, in three dimensions, the wave function has only two choices: It is positive for $\alpha = 0$, i.e., $\psi(r_1, r_2) = \psi(r_2, r_1)$ (Bose statistics) or negative for $\alpha = 1$, i.e., $\psi(r_1, r_2) = -\psi(r_2, r_1)$ (Fermi statistics)[42].

Interestingly, in a *strictly* two-dimensional system, the third dimension being absent, the above arguments do not hold. For example, it is not possible to deform continuously the path (i) to the path (ii) without the particles going through each other which is not allowed. In two dimensions, the two paths (i) and (ii) are topologically distinct, and

---

[41]Here we have followed the arguments from, Anyons, *Quantum Mechanics of Particles with Fractional Statistics*, by Alberto Lerda (Springer 1992).
[42]For simplicity, we have chosen the spins to be all polarized for fermions.

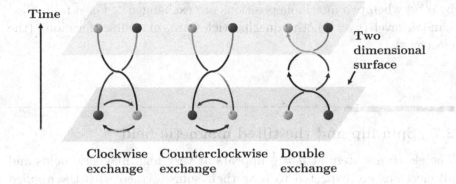

FIGURE 2.10
Braiding anyons: clockwise and counterclockwise, and the double exchange.

the phase factor for the double exchange is no longer unity. In their pioneering work, Leinaas and Myrheim[43] pointed out that in two dimensions, the statistics no longer favors one or the other choice of the wave function. The statistical parameter α is now completely arbitrary and not a multiple of π any more, but the whole range of the values of α that interpolates between bosons and fermions are now possible. Particles in two dimensions are called anyons[44]. In two dimensions, it is not enough to specify the initial and final configurations to characterize a system. Imagine the two particles being 'pinned' down and then exchanged around each other. One has to specify how their trajectories wind or 'braid' around each other (Fig. 2.10). The time evolution of the particles now becomes important.

Halperin[45] and Arovas et al.,[46] reported that Laughlin quasiparticles close to the filling factor $\nu = 1/q$, obey fractional statistics ($\alpha = 1/q$, q being an odd integer). The quasihole wave function is to be multiplied

---

[43]On the theory of identical particles, by J.M. Leinaas and J. Myrheim, *Il Nuovo Cimento B* **37**, 1 (1977).

[44]Quantum Mechanics of Fractional-Spin particles, by F. Wilczek, *Phys. Rev. Lett*, **49**, 957 (1982),

[45]Statistics of quasiparticles and the hierarchy of fractional Quantized Hall states, by B.I. Halperin, *Phys. Rev. Lett.* **52**, 1583 (1984); **52**, 2390(E) (1984).

[46]Fractional statistics and the quantum Hall effect, by D. Arovas, J.R. Schrieffer and F. Wilczek, *Phys. Rev. Lett.*, **53**, 722 (1984).

by $e^{i\pi/q}$ when two quasihole positions are exchanged[47]. For a fully filled Landau level ($q = 1$), the quasiparticles are, of course, fermions (the electrons).

2.7   Spin flip and the tilted magnetic field

The electron system at $\nu = \frac{1}{q}$ appears at very high magnetic fields and all electrons are expected to have their spins aligned with the applied magnetic field. One can therefore safely ignore the spin degree of freedom in the theory of the FQHE. Laughlin's wave function describes a fully spin polarized system, as expected at that filling factor. Soon after Laughlin proposed his theory for the effect, Halperin[48] in 1983, raised the possibility that there might be spin unpolarized states for filling factors other than $\nu = \frac{1}{q}$. In particular, he proposed that a Laughlin-like but *spin-unpolarized* state can be constructed at $\nu = 2/(p + q)$ with p and q being integers, as

$$\psi = \prod_{j<k}(z_j - z_k)^q \prod_{\alpha<\beta}(z_\alpha - z_\beta)^q \prod_{j,\alpha}(z_j - z_\alpha)^p \prod_j e^{-|z_j|^2/4\ell_0^2}$$

$$\times \prod_\alpha e^{-|z_\alpha|^2/4\ell_0^2},$$

where the Roman and Greek indices correspond to electrons with two different spin orientations. The structure of the wave function is similar to that of Laughlin, except that the third term on the right is the mixed term. The filling factor corresponding to this wave function is $\nu = 2/(p + q)$. The ground state properties of this mixed-spin state was reported for the first time for $q = 3, p = 2$, i.e., for $\nu = \frac{2}{5}$ by Chakraborty and Zhang[49]. Evaluation of the pair-correlation functions and the ground state energy followed a generalization of the

---

[47]It should however be noted that the relevant many-body wave function of the system obeys the Fermi statistics for the electrons (because electrons live in three dimensions)! The fractional statistics will appear only when the two quasiparticle positions are exchanged in a very slow [over the timescale longer than $\hbar/\left(e^2/\epsilon\ell_0\right)$] process such that the many-particle wave function evolves 'adiabatically', avoiding transitions to excited states.

[48]Theory of the quantized Hall conductance, by B.I. Halperin, *Helv. Phys. Acta* **56**, 75 (1983).

[49]Role of reversed spins in the correlated ground state for the fractional quantum Hall effect, by T. Chakraborty and F.C. Zhang, *Phys. Rev. B* **29**, 7032 (R) (1984).

techniques used for the OCP to a two-component plasma system[50]. With this work, a very intriguing possibility to observe a spin-reversed FQHE ground state and spin-reversed excitations[51] at various filling factors was thereby established. These theoretical works were subsequently confirmed by several experimental groups. A brief description of the experimental observations of the spin unpolarized states is given below.

Experimental techniques to observe the spin-reversed fractional quantum Hall states involve *tilting* the magnetic field from the direction perpendicular to the electron plane. In this situation, the Coulomb contribution to the ground state energy depends on the perpendicular component of the field $B_\perp$ but the Zeeman energy that depends on the spin orientation, in turn, depends on the total field[52] B_{total}. Tilted-field experiment in the FQHE was pioneered by Haug et al.[53], who pointed out that in a tilted field the electron-hole symmetry (i.e., the filling factors ν and $1 - \nu$ being equivalent: one for electrons while the other is for holes) that is obeyed by the Laughlin state, in now broken and $\nu = \frac{1}{3}$ and $\nu = \frac{2}{3}$ states are no longer equivalent. This is an indication of the presence of different spin polarizations in the system.

Clark et al.[54] probed spin configurations of various fractional states in a tilted field and confirmed several of our theoretical predictions of

---

[50]Variational theory of binary boson mixture at T = 0 K, by T. Chakraborty, *Phys. Rev. B* **25**, 3177 (1982); Structure of binary boson mixtures at T = 0 K, by T. Chakraborty, *Phys. Rev. B* **26**, 6131 (1982).

[51]Ground state of two-dimensional electrons and the reversed spins in the fractional quantum Hall effect, by F.C. Zhang and T. Chakraborty, *Phys. Rev. B* **30**, 7320 (R) (1984); Elementary excitations in the fractional quantum Hall effect and the spin-reversed quasiparticles, by T. Chakraborty, P. Pietiläinen, and F.C. Zhang, *Phys. Rev. Lett.* **57**, 130 (1986); Spin-dependent fractional QHE states in the N = 0 Landau level, by P.A. Maksym, *J. Phys.: Condens. Matter* **1**, L6299 (1989); Spin-reversed ground state and energy gap in the fractional quantum Hall effect, by T. Chakraborty, *Surf. Sci.* **229**, 16 (1990).

[52]Fractional quantum Hall effect in tilted magnetic fields, by T. Chakraborty, and P. Pietiläinen, *Phys. Rev. B* **39**, 7971 (1989); Subband-Landau-level coupling in the fractional quantum Hall effect in tilted magnetic fields, by V. Halonen, P. Pietiläinen, and T. Chakraborty, *Phys. Rev. B* **41**, 10202 (1990).

[53]Fractional quantum Hall effect in tilted magnetic fields, by R.J. Haug, K. v. Klitzing, R.J. Nicholas, J.C. Maan, and G. Weimann, *Phys. Rev. B* **36**, 4528 (R) (1987).

[54]Fractional quantum Hall effect in a spin, by R. Clark and P. Maksym, *Physics World* **2**, 39 (1989); Spin configurations and quasiparticle fractional charge of fractional-quantum-Hall-effect ground states in the N = 0 Landau level, by R.G. Clark et al., *Phys. Rev. Lett.* **62**, 1536 (1989).

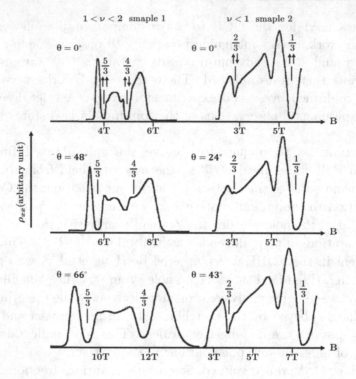

FIGURE 2.11
Effect of tilted field on the longitudinal resistance for various filling factors versus the magnetic field.

spin-reversed states. A detailed account of this aspect of the FQHE has been presented earlier by the present author[55]. Briefly, Clark et al. probed the spin configurations of several fractional filling factors, for the first time, by tilting the field B at an angle θ to the sample plane normal (increasing its absolute value at fixed density n and the perpendicular component $B_\perp$), and also by increasing the density of the system (at $\theta = 0°$) so that the fraction occurs at higher $B_\perp$. They observed that, with increasing tilt angle, dramatic changes occur in the $\rho_{xx} = R_{xx}$ minima at various filling factors. They observed that when θ and hence the total field is increased, both $\frac{2}{3}$ and $\frac{4}{3}$ states slowly disappear at first and then reappear. In contrast, the longitudinal resistance minima for $\frac{1}{3}$ and $\frac{5}{3}$ remain essentially unaltered with increasing tilt

---

[55]Electron spin transitions in quantum Hall systems, by T. Chakraborty, *Adv. Phys.* **49**, 959 (2000).

angle (Fig. 2.11). In fact, the theoretical work predicted that the $\frac{2}{3}$ and $\frac{4}{3}$ should be unpolarized, which changes to fully spin polarized when the magnetic field is increased. The Laughlin state at $\frac{1}{3}$ is that for a fully spin polarized state. The experiments of Clark et al. established the theoretical predictions by us on the role of reversed spin in the FQHE on a firm footing. There were several other experiments reported in the literature providing supporting evidence for the spin structure of the FQHE[56].

2.8 Anatomy of the Laughlin state

As we discussed in the previous section, it is now well established that the Laughlin state correctly describes the strongest $\frac{1}{3}$-FQHE observed in experiments, and it is the prototype model for the other FQHE states. However, even after the astounding success of this wave function and the subsequent intense research that spanned for almost four decades, a fundamental question has remained unanswered: What is the origin of incompressibility and how does Laughlin's wave function successfully describe the incompressibility of the $\frac{1}{3}$ state. It has been pointed out by Haldane that none of the existing theories of FQHE, be it the topological quantum field theories, or filling of effective Landau levels by composite fermions, provide a quantitative proof of the origin of incompressibility of the FQHE states. Haldane has recently made an effort to generalize the Laughlin state to answer that question, by introducing the geometry fluctuations in the state[57]. The mathematical steps of this approach are quite cumbersome (as opposed to the simple and elegant approach of Laughlin), and will not be discussed in detail.

Briefly, in dealing with the FQHE, Haldane introduced several 'metrics' in the problem. First, the 2D system of interacting electrons is described by an 'anisotropic' mass tensor for the single-electron state (see Box 2.7), as opposed to the circularly symmetric case adapted by Laughlin. That results in a 'Galilean metric' that is unimodular (area

---

[56]Evidence for a phase transition in the fractional quantum Hall effect, by J.P. Eisenstein, H.L. Störmer, L. Pfeiffer, and K.W. West, *Phys. Rev. Lett.* **62**, 1540 (1989); Fractional quantum Hall effect at $\nu = \frac{2}{3}$ and $\frac{3}{5}$ in tilted magnetic fields, by L.W. Engel, S.W. Hwang, T. Sajoto, D.C. Tsui, and M. Shayegan, *Phys. Rev. B* **45**, 3418 (1992).

[57]Geometrical description of the fractional quantum Hall effect, by F.D.M. Haldane, *Phys. Rev. Lett.* **107**, 116801 (2011).

preserving). This metric defines the shape of the Landau level orbitals. The second metric arises from the dielectric tensor of the semiconducting material and defines the shape of the equipotential contours around an electron. Rotational invariance in the Laughlin approach means that these two metrics are congruent. However, they might be different from one another in actual systems and hence the lifting of rotational invariance. As explained in Box. 2.7, anisotropy plays an important role in the geometric approach to the Laughlin state. In this respect, a recently isolated single layer of black phosphorus (phosphorene), discussed in Sec. 5.7, with its anisotropic band structure might be a suitable system for exploration.

Box: 2.7 Single-particle Hamiltonian and mass anisotropy:

Using the Einstein summation notation, the Hamiltonian for a single-particle moving in a 2D xy-plane and subjected to a magnetic field $B^z = \epsilon^{ab}\partial_a A_b$, where $\epsilon_{ab} = \epsilon^{ab}$ is the 2D antisymmetric Levi-Civita symbol, is given by

$$\mathcal{H} = \tfrac{1}{2}\left(m^{-1}\right)^{ab}\pi_a\pi_b.$$

Here m_{ab} is the cyclotron effective mass tensor and the kinematic momentum π_a is given by $\pi_a = p_a - eA_a$. The mass tensor is

$$M = (m_{ab}) = \begin{pmatrix} m_{xx} & m_{xy} \\ m_{xy} & m_{yy} \end{pmatrix}.$$

Since M is symmetric, there exists an orthogonal transformation $\mathcal{O}$ which takes it to the diagonal form:

$$D = m\begin{pmatrix} \alpha & 0 \\ 0 & 1/\alpha \end{pmatrix} = m\Delta, \quad \alpha = \sqrt{m_{xx}/m_{yy}}, \quad m = \sqrt{m_{xx}m_{yy}}$$

where $\Delta = \begin{pmatrix} \alpha & 0 \\ 0 & 1/\alpha \end{pmatrix}$. As a consequence, $M = m\mathcal{O}\Delta\mathcal{O}^{\mathrm{T}} = mG^{-1}$, where $G = \mathcal{O}\Delta^{-1}\mathcal{O}^{\mathrm{T}}$, $\Delta^{-1} = \begin{pmatrix} 1/\alpha & 0 \\ 0 & \alpha \end{pmatrix}$. Obviously, G is symmetric: $G^{\mathrm{T}} = \left(\mathcal{O}\Delta^{-1}\mathcal{O}\right)^{\mathrm{T}} = \mathcal{O}\Delta^{-1}\mathcal{O}^{\mathrm{T}} = G$. Using the orthogonality of the matrix $\mathcal{O}$ we also have $\det G = \det\left(\mathcal{O}\Delta^{-1}\mathcal{O}^{\mathrm{T}}\right) =$

$\det\left(\mathcal{O}^{\mathrm{T}}\mathcal{O}\Delta^{-1}\right) = \det\Delta^{-1} = 1$, i.e., the matrix G is *unimodular*. We now define the metric tensor, $g^{ab} = (G)_{ab}$ and its inverse g_{ab} via the relation, $g^{ab}g_{bc} = \delta^a_c$. The single-particle Hamiltonian is then

$$\mathcal{H} = \tfrac{1}{2m}g^{ab}\Pi_a\Pi_b$$

for the anisotropic system described by Haldane. The Galilean metric (g_{ab}) and other metrics introduced by Haldane play crucial roles in the geometrical description of the FQHE.

The basic reason for introducing anisotropy in the FQHE is that, anisotropy helps one to vary the metrics to unravel the geometric description of the FQHE. As shown below, one possible way to break the rotational symmetry is to apply a tilted magnetic field. Haldane introduced the interaction metric in the Laughlin state as a variational parameter that minimizes the correlation energy. Interestingly, this is the basic tenet of the original Jastrow variational theory[58], where the Jastrow function contains variational parameters that minimize the interaction energy. Incompressibility of the Laughlin state results from fluctuation of this quantum geometry.

Box 2.8 Additional anisotropy:

Assuming that the only anisotropy in the system is due to the effective mass, we can always choose our coordinate frame such that its axes coincide with the principal axes of the equimass ellipses. In this frame, the metric $G = \begin{pmatrix} 1/\alpha & 0 \\ 0 & \alpha \end{pmatrix}$ is diagonal. However, if there is an additional anisotropy making an azimuthal angle ϕ, we could still, as a first step, orient the frame along the principal axes of the mass ellipses resulting in a diagonal metric. As a second step,

[58]Hypernetted-Chain Theory of Matter at Zero Temperature, by J.G. Zabolitzky, in: J.W. Negele, E. Vogt (eds) *Advances in Nuclear Physics. Advances in the Physics of Particles and Nuclei*, vol 12. Springer, Boston, MA (1981).

we could rotate the frame by the transformation

$$x' = x \cos\phi + y \sin\phi$$
$$y' = -x \sin\phi + y \cos\phi$$

to make the new x'-axis to coincide with the anisotropy. The matrix

$$R = \begin{pmatrix} \cos\phi & \sin\phi \\ -\sin\phi & \cos\phi \end{pmatrix}$$

of the rotation is obviously orthogonal, i.e., $R^{\mathrm{T}}R = 1$. Under the rotation

$$G' = R^{\mathrm{T}}GR = \begin{pmatrix} \cos\phi & -\sin\phi \\ \sin\phi & \cos\phi \end{pmatrix} \begin{pmatrix} 1/\alpha & 0 \\ 0 & \alpha \end{pmatrix} \begin{pmatrix} \cos\phi & \sin\phi \\ -\sin\phi & \cos\phi \end{pmatrix}$$

which, after some steps leads to

$$G' = \begin{pmatrix} \cosh 2\theta + \sinh 2\theta \cos 2\phi & \sinh 2\theta \sin 2\phi \\ \sinh 2\theta \sin 2\phi & \cosh 2\theta - \sinh 2\theta \cos 2\phi \end{pmatrix},$$

where $\cosh 2\theta = \frac{1}{2}(\alpha + 1/\alpha)$. This is equivalent to the metric derived by Haldane et al. in the case of a tilted magnetic field from the direction perpendicular to the electron plane.

One interesting outcome of this rather complex theory is that the mass anisotropy introduced in this scheme can indeed be shown to be equivalent to the tilted-field effect on the lowest Landau level single-particle states. The metric associated with the tilted field, as shown by Haldane et al.[59], is mathematically equivalent to the metric tensor due to variation of the anisotropic mass tensor as presented in Box 2.8.

As it now stands, the results in Box 2.8 are more of a mathematical curiosity than anything profound about the Laughlin state. It seems that in this approach, to paraphrase the Scots man of letters Andrew Lang (1844–1912), anisotropy has been used 'as a drunken man uses lamp-posts – for support rather than illumination'. As we have discussed

---

[59]Model anisotropic quantum Hall states, by R.-Z. Qiu, et al., *Phys. Rev. B* **85**, 115308 (2012); Band mass anisotropy and the intrinsic metric of fractional quantum Hall systems, by Bo Yang, et al., *Phys. Rev. B* **85**, 165318 (2012).

above, the tilted-field has been merely a tool to resolve the spin polarization of the FQHE states. However, the anisotropy of the quantum Hall liquid may display even more useful properties[60], relevant to the Laughlin function. A better understanding of the Laughlin state will also help in unravelling the intricate behavior of Laughlin quasiparticles and quasiholes. Clearly, more work will be required for this interesting approach to come to fruition[61].

In order to understand the success of Laughlin's wave function in describing the incompressible state, one could also proceed along the following route: We have seen in Sec. 2.5 that the Laughlin wave function has close ties to the one-component classical plasma. Therefore, it would be interesting to understand how the compressibility behaves in the plasma. It has been reported in the literature that the compressibility takes on negative values when the plasma coupling parameter Γ (defined in Box 2.3, and related to the Laughlin wave function as $\Gamma = 2q$ for $q = 3$) *exceeds* the critical value[62] $\Gamma_c \sim 3$. It should be pointed out that the compressibility, being the second derivative of the internal energy with respect to the particle density strongly depends on the accuracy of the method used to evaluate the energy. In fact, the value of the compressibility in the plasma strongly depends on the accuracy of the theoretical method employed. Therefore, evaluation of Γ_c is still a theoretical challenge.

2.9 Laughlin state from the East

In 2011, the fractional quantum Hall effect (FQHE) at $\nu = \frac{1}{3}$ was observed in a totally new system: in the two-dimensional electron gas (2DEG) created at the oxide interfaces, in particular, in MgZnO/ZnO heterostructures[63]. This is a very exciting finding for various reasons.

---

[60]Incompressible states of dirac fermions in graphene with anisotropic interactions, by V.M. Apalkov and T. Chakraborty, *Solid State Commun.* **177**, 128 (2014).

[61]The origin of holomorphic states in Landau levels from non-commutative geometry and a new formula for their overlaps on the torus, by F.D.M. Haldane, *J. Math. Phys.* **59**, 081901 (2018).

[62]On the compressibility of a one-component plasma, by M. Baus, *J. Phys. A: Math. Gen.* **11**, 2451 (1978).

[63]Insulating phase of a two-dimensional electron gas in $Mg_xZn_{1-x}O$/ZnO heterostructures below $\nu = \frac{1}{3}$, by Y. Kozuka, et al., *Phys. Rev. B* **84**, 033304 (2011).

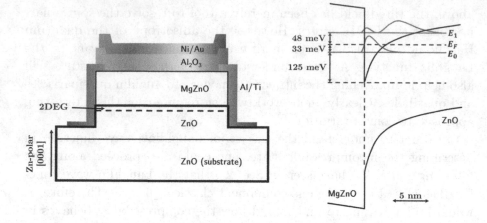

FIGURE 2.12
Schematic cross-sections of a sample. The 2DEG is located near the MgZnO/ZnO interface in a Zn-polar ZnO substrate.

Until now, these studies were mostly limited to GaAs systems (other than graphene, discussed in Chapter 5) due to the requirement of extremely clean 2DEGs, that is essential for observation of the FQHE. That monopoly has now been broken with the advent of 2D electrons with high mobility at the oxide interfaces (Fig. 2.12).

ZnO is well known as an electronic semiconductor with a large band gap (3.37 eV at room temperature). Interestingly, formation of the 2DEG in this system does not require any external field or impurity doping: the 2DEG is naturally present when the interface is formed[64]. The new system[65] possess several very exciting properties that might help in further

---

[64]Challenges and opportunities of ZnO-related single crystalline heterostructures, by Y. Kozuka, A. Tsukazaki and M. Kawasaki, *Appl. Phys. Rev.* **1**, 011303 (2014).

[65]Zinc oxide (ZnO) is the white powder commonly used today as the UV-light absorbing component of the sun lotions. However, the origin of the use of Zinc goes back to more than 2,000 years in India, where it was extensively used in producing brass, an alloy of copper and zinc. In fact, the alloy containing up to 12% of zinc can literally shine like gold, after a good polish! In that respect, the indian alchemists were closest to achieving gold in ancient times. Brass has been traditionally used for centuries in the furniture and other accessories in Hindu and Buddhist temples. For more historical facts about zinc, please see, Rasa-Ratna-Samuccaya and mineral processing state-of-the-art in the 13th century AD India, by A.K. Biswas, *Indian J. Hist. Sci.* **22**, 22 (1987); Zinc production in medieval India, by P.T. Craddock, L.K. Gurjar, and K.T.M. Hegde, *World Archaeology*, **15**, 211 (1983); Zinc – The metal from the East, by F. Habashi, *Bull. Can. Inst. Min. Met.* **94**, 71 (2001).

explorations of quantum phenomena in quantum confined systems, perhaps from entirely different perspectives[66]. The 2D electron system in ZnO has some unique properties as compared to those in GaAs. The electron effective mass in ZnO is 0.29 m_e, while for GaAs it is 0.069 m_e. Here m_e is the free-electron mass. The kinetic energy is therefore reduced in ZnO, thus making the interaction energy more dominant. Further, the g-factor of GaAs is -0.44 while for ZnO it is 4.3. While in GaAs, the Zeeman energy is seventy times smaller than the cyclotron energy, in ZnO, that ratio is 0.94, closer to unity. This can lead to reversal in the order in which orbital Landau levels (LLs) and spin states are filled, at certain filling fractions. The dielectric constant of GaAs is 13 while in ZnO it is 8.5. For a density of $n \sim 10^{11}$ cm^{-2}, the ratio of Coulomb to the Fermi energy is $1.8/\sqrt{n}$, while it is $11.9/\sqrt{n}$ for ZnO. Electrons in ZnO are therefore strongly correlated. In the case of GaAs-based electron gas, the Landau level gap is large compared to that for the Coulomb interaction. In a ZnO heterostructure, the LL gap is very small. This has resulted in several unexpected findings in the quantum Hall effects[67]. It is a promising candidate for quantum devices with high electron mobility and strong electron-electron correlations that might lead to more emergent novel phenomena in the future, and offer the exciting possibility of all-oxide electronic devices[68]. Some unusual effects of nanostructures created with ZnO are discussed in Sec. 4.7. Incidentally, our encounter with the quantum Hall effects, both integral and fractional, is still not over yet, as those effects appear as new 'avatars' in Chapter 5!

---

[66]Observation of the fractional quantum Hall effect in an oxide, by A. Tsukazaki, et al., *Nat. Materials* **9**, 889 (2010).

[67]Phase transitions at $\nu = \frac{5}{2}$ in ZnO-based heterostructures, by J. Falson, *Physica E* **110**, 49 (2019); A review of the quantum Hall effects in MgZnO/ZnO heterostructures, by J. Falson and M. Kawasaki, *Rep. Prog. Phys.* **81**, 056501 (2018); Missing fractional quantum Hall states in ZnO, by W. Luo and T. Chakraborty, *Phys. Rev. B* **93**, 161103(R) (2016); Tilt-induced phase transitions in even-denominator fractional quantum Hall states at the ZnO interface, by W. Luo and T. Chakraborty, *Phys. Rev. B* **94**, 161101(R) (2016); Pfaffian state in an electron gas with small Landau level gaps, by W. Luo and T. Chakraborty, *Phys. Rev. B* **96**, 081108(R) (2017).

[68]Emergent phenomena at oxide interfaces, by H.Y. Hwang, et al., *Nat. Materials* **11**, 103 (2012).

3

Quantum dots: In the abyss of no dimensions

A satirical novella, and yet an amazing geometrical allegory in the Victorian era, *Flatland: A Romance of Many Dimensions* by Edwin A. Abbott describes the journey of the narrator A Square from his limited world of two dimensions, guided by a three-dimensional sphere to the 'Pointland' as follows: 'Look yonder,' said my Guide, 'in Flatland thou hast lived; ... thou hast soared with me to the heights of Spaceland; now, in order to complete the range of thy experience, I conduct thee downward to the lowest depth of existence, even to the realm of Pointland, the Abyss of No dimensions'.

As we have learned in Chapter 2, confinement of electron motion in one direction creates a two-dimensional electron system. Similarly, confinement in all three directions creates the zero-dimensional electron systems: the quantum dots. These systems are popularly known as 'artificial atoms'[1]. The reason for this unusual nomenclature is because the confinement of electrons in all three spatial directions results in a discrete energy spectrum, just as in atoms. A major advantage of studying quantum dots is that their shape, size, and the number of electrons can be controlled very precisely by experiment. In stark contrast to the description of 'pointland' by the sphere, these zero-dimensional objects are far from being abysmal but instead are filled with rich physics and great potentials to create incredible devices[2]. These systems are thought to have vast potential for future technological applications[3] not least in biology[4]. One prominent example is quantum cryptography, where

---

[1]To the best of our knowledge, this popular name was introduced in the literature by P. Maksym and T. Chakraborty in *Phys. Rev. Lett.* **65**, 108 (1990). Other more appropriate names were 'designer atoms' by M. Reed, *Sci. Amer.* **268**, 98 (1993), and 'Superatom' by H. Watanabe and T. Inoshita, *Optoelectron. Dev. Technol.* **1**, 33 (1986).

[2]*Quantum Materials*, edited by D. Heitmann (Springer 2010)

[3]Quantum dots go on display, by K. Bourzac, *Nature* **493**, 283 (2013).

[4]*Quantum Dots: Applications in Biology*, edited by M.P. Bruchez and C.Z. Holz (Humana Press, New Jersey 2007).

DOI: 10.1201/9781003090908-3

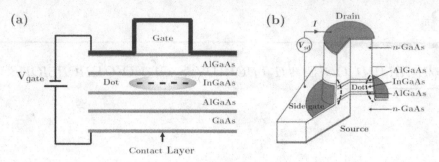

FIGURE 3.1
Schematic illustration of typical electrostatic quantum dots. In (b) the dot is located between the two AlGaAs tunnel barriers. A negative voltage applied to the side gate squeezes the dot thus reducing the effective diameter of the dot (dashed curves). The dot in (a) is generally used for optical spectroscopy measurements, while the dot in (b) is used for transport measurements.

quantum dots are used to make single photon emitters[5] and single photon detectors[6]. Quantum cryptography, in turn, is ushering in what is called the 'quantum internet'[7] (see Sec. 3.8).

The quantum dots are usually fabricated by applying a lateral confinement to a two- dimensional electron system. Typically, the two-dimensional system is in the heterojunctions[8], as described in Chapter 2, or in a quantum well[9] and the dot is created by etching away

---

[5]Single-photon source – an introduction, by S. Scheek, *J. Mod. Optics* **56**, 141 (2009); Deterministic and electrically tunable bright single-photon source, by A.L. Nowak, et al., *Nature Comm.* **5**, 3240 (2014); A quantum dot single-photon source, by C. Becher, et al., *Physica* E **13**, 412 (2002); A quantum dot single photon source driven by resonant electrical injection, by M.J. Conterio, et al., *Appl. Phys. Lett.* **103**, 162108 (2013).

[6]Detection of single photons using a field-effect transistor gated by a layer of quantum dots, by A.J. Shields, et al., *Appl. Phys. Lett.* **76**, 3673 (2000); Single photon detection with a quantum dot transistor, by A.J. shields, et al., *Jpn. J. Appl. Phys.* **40**, 2058 (2001).

[7]Satellite-relayed intercontinental quantum network, by S-K Liao, et al., *Phys. Rev. Lett.* **120**, 030501 (2018).

[8]Zeeman bifurcation of quantum-dot spectra, by W. Hansen, et al., *Phys. Rev. Lett.* **62**, 2168 (1989).

[9]Observation of discrete electronic states in a zero-dimensional semiconductor nanostructure, by M.A. Reed, et al., *Phys. Rev. Lett.* **60**, 535 (1988).

part of the surface, but a more common alternative is to use a modulated gate electrode[10]. A small region of the gate is in the form of a cap which is further away from the electron layer than the rest. When a negative voltage is applied to the gate, the electrons in the regions closest to the gate are fully depleted but a few electrons remain in the region under the cap and form the quantum dot [Fig. 3.1 (a)]. Alternatively, in a pillar structure [Fig. 3.1 (b)], the dot is located in the center of the pillar containing electrons. Here the diameter of the quantum dot is a few hundred nanometers and the thickness is about 10 nm. As shown, the dot is sandwiched between two non-conducting barrier layers, while separated from conducting materials above and below. Application of a negative voltage to a metal gate aroud the pillar depletes the number of electrons in the dot[11]. Quantum dots created in this type of structures are used for transport measurements. Another very important fabrication technique involves self-organized growth of InAs quantum dots on GaAs substrates[12]. These nanoscale islands form when a thin layer of InAs is deposited on a substrate of GaAs. Under appropriate conditions, these islands can be filled with electrons. Self-assembled quantum dots have been recently used in single-electron charge sensing[13].

We begin by discussing some historical issues that are intimately related to the physics of quantum dots.

3.1 Landau versus Fock

In Chapter 2, we mentioned that the discrete energy levels of an electron in a strong magnetic field are known as Landau levels. They were named

[10]The spectroscopy of quantum dot arrays, by D. Heitmann and J. Kotthaus, *Phys. Today* **46**, 56 (1993).

[11]Few-electron quantum dots, by L.P. Kouwenhoven, D.G. Austing, and S. Tarucha, *Rep. Prog. Phys.* **64**, 701 (2001).

[12]Self-assembled quantum dots, edited by Z.M. Wang (Springer, 2008); Optical spectroscopy of self-assembled quantum dots, by D. Mowbray and J. Finley, in *Nano-Physics and Bio-Electronics: A New Odyssey*, edited by T. Chakraborty, F. Peeters, and U. Sivan (Elsevier, 2002).

[13]Single-electron charge sensing in self-assembled quantum dots, by H. Kiyama, et al., *Sci. Rep.* **8**, 13188 (2018); and references therein.

after the famous soviet physicist L. Landau[14]. Unfortunately, only a few people knew that another soviet physicist V. Fock published a much more general theory two years prior[15] to Landau's work, where the 'Landau levels' were only a special case! Later, Darwin[16] also reported similar studies. Apparently, neither Landau, nor Darwin was aware of Fock's paper! In that brief publication, Fock solved the problem of an electron subjected to both a uniform magnetic field *and* a harmonic oscillator potential of the form $\frac{1}{2}m\omega_0^2 r^2$, where ω_0 is the potential strength. In this case, the energies of an electron are: $E_{nl} = (2n + |l| + 1)\hbar\Omega - \frac{1}{2}l\hbar\omega_c$ (without the Zeeman energy, i.e., the energy difference between the two spin states). Here $l = 0, \pm1, \pm2, \ldots$ and $n = 0, 1, 2, \ldots$, respectively, are angular momentum and radial quantum numbers, $\Omega^2 = \left(\frac{1}{4}\omega_c^2 + \omega_0^2\right)$, $\omega_c = eB/m$, and $\hbar\omega_0$ is the confinement energy. In the limit of zero confinement ($\omega_0 \to 0$), the energy levels are $E_{nl} = \left(\mathcal{N} + \frac{1}{2}\right)\hbar\omega_c$, where $\mathcal{N} = n + (|l| - l)/2$ is the Fock-Darwin level (FDL) index. This is shown in Fig. 3.2 for a Gallium Arsenide (GaAs) semiconductor quantum dot[17].

In the absence of a magnetic field ($B = 0$), the levels are equally spaced (just as the 'particle in a box' problem in textbook quantum mechanics) and as the field increases a complicated series of level crossings occur. Interestingly, in the limit $B \to \infty$ the levels coalesce into the Landau levels. However, there is significant broadening even at 12 Tesla. The degeneracy of the LLs discussed in Chapter 2 are now lifted by the confinement potential[18].

Experimental investigations of quantum dots include electron transport through the dots, charging of dots and optical spectroscopy. Electron tunneling from the source to the quantum dot and from the dot to the drain crucially depends on the barrier heights. For a high-tunneling barrier, tunneling into and from the dot is weak and only an integer number N of electrons can be in the dot. Adding an extra electron requires extra energy because of the Coulomb repulsion between the electrons. Both charging of quantum dots and transport through the dots

[14]Diamagnetismus der Metalle, *Zeit. f. Physik* **64**, 629 (1930).

[15]Bemerkung zur Quantelung des harmonischen Oszillators im Magnetfeld, by V. Fock, *Zeit. f. Phys.* **47**, 446 (1928).

[16]The Diamagnetism of the Free Electron, by C.G. Darwin, *Math. Proc. Cambridge Phil. Soc.* **27**, 86 (1931).

[17]*Physics of the Artificial Atoms: Quantum Dots in a Magnetic Field*, by T. Chakraborty, Comm. Condens. Matter Phys. **16**, 35 (1992); Transport phenomena in mesoscopic systems, eds. H. Fukuyama and T. Ando (Springer, 1992).

[18]*Quantum Dots: A Survey of the Properties of Artificial Atoms*, by T. Chakraborty (North-Holland, Amsterdam 1999).

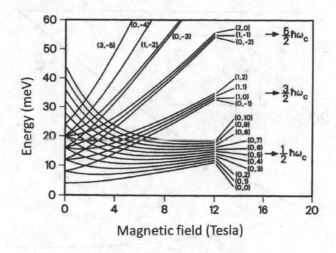

FIGURE 3.2

Single-electron energy levels for a parabolically confined quantum dot in a magnetic field (in units of Tesla). The levels are indicated by their quantum numbers (n, l). The confinement energy for this dot is $\hbar\omega_0 = 4$ meV.

invoke this Coulomb blockade to control tunneling of electrons into and out of a dot[19]. In each case the dot is made such that electrons can tunnel in and out of it from a reservoir with electrochemical potential $\mu = E(N) - E(N-1)$, where $E(N)$ is the ground state energy of the N-electron dot (see Footnotes 17, 18). Therefore, μ is the energy required to add the $(N+1)$th electron to the dot. At low temperatures, the number of electrons is fixed at N if $\mu(N+1) > \mu > \mu(N)$. An electron can, however, tunnel into the dot when $\mu(N+1) = \mu$. This resonance condition can be met for different values of N by changing the gate voltage.

[19]Artificial atoms, by M. Kastner, *Phys. Today* **46**, 24 (1993).

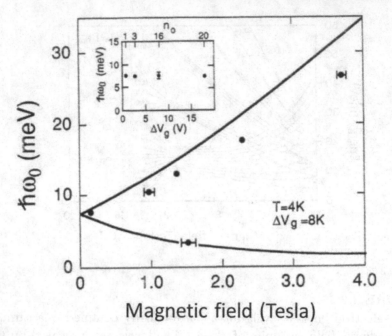

FIGURE 3.3
Resonance position of quantum dots. The solid lines are obtained from
the allowed excitation energies of the Fock-Darwin energy spectrum.
Independence of the energies on the gate voltage and the electron number
is shown as inset.

3.2   A tale of artificial atoms

In 1989, Dr. Ulrich Merkt at the University of Hamburg was puzzled by
a very surprising result. He performed the first magneto-optical experiments on quantum dots. The spectroscopy of the dots was carried out
with linearly polarized far-infrared (FIR) laser light, and the results[20] are
shown in Fig. 3.3. Here the good news is that the data closely follows the
allowed transitions in the Fock-Darwin energy levels, $\Delta E_{\pm} = \hbar\Omega \pm \frac{1}{2}\omega_c$
(see Footnotes 10, 20). The upper mode ΔE_+ in Fig. 3.3 approaches the

---

[20]Spectroscopy of electronic states in InSb quantum dots, by Ch. Sikorski and U.
Merkt, *Phys. Rev. Lett.* **62**, 2164 (1989).

cyclotron energy $\hbar\omega_c$ (transitions from one Landau level to another) in the limit of a large magnetic field, while the lower mode ΔE_- approaches zero in that limit. The surprising result is that this seems to be true for quantum dots containing upto 20 electrons in the dots that he studied. Clearly, the electrons at such close quarters are seemingly oblivious to the repulsive Coulomb interaction among the electrons, which should be rather strong in that situation! Maksym and Chakraborty[21] resolved this puzzle by pointing out that *only* for a parabolic confinement the interacting many-electron Hamiltonian separates into terms which are functions of the center-of-mass (CM) and relative coordinates. Due to some cancellations, the CM term was found to have exactly the same energy eigenvalues as those of a single confined electron. The FIR radiation only excites the CM but does not affect the relative motion. That is the reason the FIR experiments in quantum dots display only features at the single-electron level as long as the confinement potential is parabolic. Even slight deviations from this special case of a *strictly* parabolic confinement introduces rich structure in the FIR spectra[22] with the appearance of anticrossings and additional modes due to electron-electron interactions. The interaction effects can only be probed by either deliberately engineering the dots so that the CM and relative motions are coupled or by measuring the thermodynamic properties of the electrons.

3.3 Portrait of a harmonic oscillator

Oscillating systems are present everywhere in Nature, and simple harmonic oscillators are the most important models in many natural phenomena. They are ubiquitous in both classical and quantum systems, and are the most studied systems in Nature. In fact, any system near a stable equilibrium is equivalent to a harmonic oscillator. Undoubtedly, the harmonic oscillator is one of the most important examples of motion

---

[21] Quantum dots in a magnetic field: Role of electron-electron interactions, by P.A. Maksym and T. Chakraborty, *Phys. Rev. Lett.* **65**, 108 (1990).

[22] Quantum-dot helium: Effects of deviations from a parabolic confinement potential, by D. Pfannkuche and R.R. Gerhardts, *Phys. Rev. B* **44**, 13132 (1991); Magneto-optical transitions and level crossings in a Coulomb-coupled pair of quantum dots, by T. Chakraborty, V. Halonen, and P. Pietiläinen, *Phys. Rev. B* **43**, 14289 (1991).

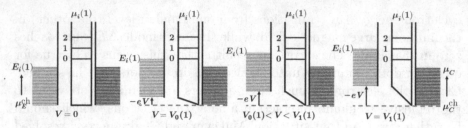

FIGURE 3.4

Energy diagram (schematic) illustrating the transport spectroscopy of single-particle spectrum of a quantum dot.

in the entire physics. It played a crucial role at the very beginning of the quantum theory[23]. Later, in quantum mechanics the zero-point energy of vibration (at absolute zero) of the quantum harmonic oscillator is entirely due to Heisenberg's uncertainty principle[24]. The reason for its popularity is that it is one of the very few exactly solvable model, in particular, in quantum mechanics. In the presence of an applied magnetic field, the Fock-Darwin levels, and the Landau levels are manifestations of the harmonic oscillator motion.

The single-electron tunneling spectroscopy on double-barrier structure provides a glimpse of the energy spectrum of confined electrons[25]. The basic idea is schematically sketched in Fig. 3.4. The crucial quantity here to understand the experimental results is the addition energy, $\mu(N)$, i.e., the energy required to add one extra electron to a $(N-1)$-electron dot. At zero bias $(V = 0)$ the dot is empty of electrons since the ground state energy $E(1)$ of the first electron $[E(1) = \mu(1)]$ lies above the chemical potential of the emitter and collector contacts $\mu^{\text{ch}}_{\text{E,C}}$. Whenever the addition energy μ_{E} exceeds $\mu_i(1) - \alpha eV$, α being the voltage to energy conversion coefficient and $\mu_i(1) = E_i(1)$, the single-electron states are available for tunneling from the emitter to the dot and one observes a step in the measured current. The bias position of these steps $V_i(1) = \left[\mu_i(1) - \mu^{\text{ch}}_{\text{E}}\right]/e\alpha$, provides a direct experimental probe

[23] *Quantum Physics of Atoms, Molecules, Solids, Nuclei, and Particles*, by R. Eisberg and R. Resnick (John Willey and Sons, 2nd edition 1985).

[24] Uncertainty principle and the zero-point energy of the harmonic oscillator, by R.A. Newing, *Nature* **136**, 395 (1935).

[25] Quantum-dot ground states in a magnetic field studied by single-electron tunneling spectroscopy on double-barrier heterostructures, by T. Schmidt, et al., *Phys. Rev. B* **51**, 5570 (R) (1995).

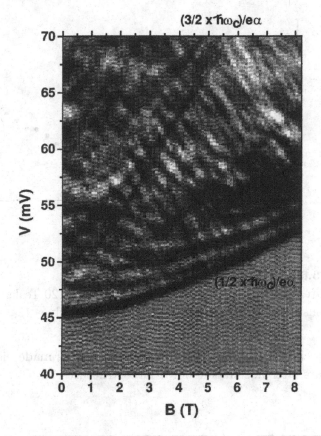

FIGURE 3.5
Differential conductance $G = dI/dV$ as a function of the magnetic field B (in Tesla) and the bias voltage. The location of the lowest Landau level energy $\left(\frac{1}{2}\hbar\omega_c\right)/e\alpha$ and the second Landau level $\left(\frac{3}{2}\hbar\omega_c\right)/e\alpha$ are indicated.

to the single-particle spectrum including the ground state. The grayscale plot of Fig. 3.5 shows the differential conductance $G = dI/dV$ as a function of the bias voltage and the magnetic field (derived from the current (I) - voltage (V) data) as obtained by Schmidt[26]. Some of the low-lying Fock-Darwin states are clearly discernible from this figure. It is quite remarkable how the energy spectrum of an *ideal* quantum

[26]*Dissertation*, by T. Schmidt, Max-Planck Institute, Stuttgart, 1997.

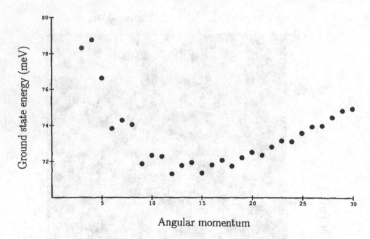

FIGURE 3.6
Ground state energy of three interacting electrons at 20 Tesla and $\hbar\omega_0 = 4$ meV versus the angular momentum quantum number J.

harmonic-oscillator model derived by Fock in 1928 is made visible some seventy years later in semiconductor nanostructures.

3.4   Magic number ground states

Quantum states of many *interacting* electrons in a quantum dot are studied usually by numerically diagonalizing the many-electron Hamiltonian[27]. Only a few electrons can be treated (albeit exactly) in this approach. An alternative approach is the current density-functional theory[28], where a much larger number of electrons can be studied. The magnetic field dependence of the ground state is particularly interesting and is relevant to many measurable properties of quantum dots.

---

[27]See Footnotes 17, 18. See also, Electronic structure of ultrasmall quantum-well boxes, by G.W. Bryant, *Phys. Rev. Lett.* **59**, 1140 (1987); Generation of Coulomb matrix elements for the 2D quantum harmonic oscillator, by M. Pons Viver, and A. Puente, *J. Math. Phys.* **60**, 081905 (2019).

[28]Current-density-functional theory of quantum dots in a magnetic field, by M. Ferconi and G. Vignale, *Phys. Rev. B* **50**, 14722(R) (1994).

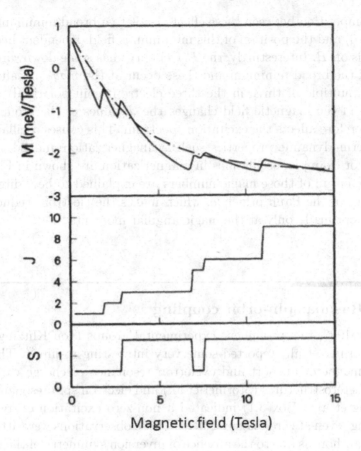

FIGURE 3.7

Magnetic field dependence of magnetization for three interacting electrons. The dashed curve corresponds to the non-interacting system. The magnetic field dependence of the total spin (S) and the total angular momentum (J) are also shown.

The Fock-Darwin states are localized on rings whose radius R can be expressed as, $R^2 \sim 2\lambda^2 (2n + |l| + 1)$ where the length parameter $\lambda^2 = \hbar/(2m\Omega)$ sets the length scale and determines the ground state angular momentum. Large angular momentum implies a large confinement energy and small angular momentum means a large repulsive energy[29].

---

[29]Magic number ground states of quantum dots in a magnetic field, by P.A. Maksym, *Physica B* **184**, 385 (1993).

The competition between these effects leads to a broad minimum $E_0(J)$ (Fig. 3.6) and the position of this minimum is field dependent because λ depends on B. Interestingly, the $E_0(J)$ curve has some downward cusps around the broad minimum and these occur at the *magic J* values that are all multiple of three in the three electron (spin polarized) case. A change in the magnetic field changes the J values of the ground state and therefore affects the excitation spectrum. This causes oscillations in the thermodynamic properties such as magnetization and heat capacity[30]. For example, oscillations in magnetization are shown in Fig. 3.7.

The origin of those magic numbers are explained to be a direct consequence of the Pauli principle, which makes the electrons reduce their energy optimally only at the magic angular momenta[31].

3.5   Rashba spin-orbit coupling

In July 1983, two prominent experimental groups (von Klitzing et al., and Störmer et al.) reported some very interesting results[32]. The combined magnetotransport and cyclotron resonance of charge carriers in GaAs-heterostructures (Störmer et al.) and electron spin resonance (von Klitzing et al.) (Box 3.1) indicated a non-zero excitation energy (spin splitting) even at zero magnetic field. Those observations were attributed by the authors as due to the absence of inversion symmetry of the roughly triangular carrier confinement potential, at the heterojunction interface (see Fig. 2.2).

Box 3.1 Cyclotron and electron spin resonances:

In the two dimensional electron gas, we have discussed about the Landau levels and in Fig. 2.7, we have shown that the Landau level separation energies to be $\hbar\omega_c$. If electromagnetic radiation

---

[30]Effect of electron-electron interactions on the magnetizations of quantum dots, by P.A. Maksym and T. Chakraborty, *Phys. Rev. B* **45**, 1947(R) (1992).

[31]See Footnote 29.

[32]Electron spin resonance on GaAs-Al$_x$Ga$_{1-x}$ heterostructures, by D. Stein, K.v. Klitzing, and G. Weimann, Phys. Rev. Lett. **51**, 130 (1983); Energy structure and quantized Hall effect of two-dimensional holes, by H.L. Störmer, et al., *Phys. Rev. Lett.* **51**, 126 (1983).

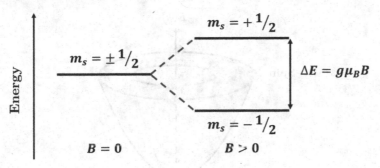

FIGURE 3.8

Energy states of an unpaired electron in an applied magnetic field B.

is incident on the system, the cyclotron motion of the electron resonantly absorbs the energy when the angular frequency of the radiation ω coincides with the cyclotron frequency ω_c. This is what is called the cyclotron resonance which is a powerful method to determine the masses of the charge carriers and to study the electronic states of the system.

An unpaired electron, in addition of having the charge, also has a spin and an associated magnetic moment due to the spin. An electron of spin $s = \frac{1}{2}$ can have the spin angular momentum quantum number of $m_s = \pm\frac{1}{2}$. In the absence of an applied magnetic field, the two values of m_s will give rise to two doubly degenerate[33] spin energy state. In an applied magnetic field, the degeneracy is no longer present and there is a splitting of the spin states (Fig. 3.8). The lower energy state (more stable one) has the magnetic moment parallel to the applied field, while the upper energy state (less stable) has its spin aligned against the magnetic field. The transition energy between the two states is $\Delta E = g\mu_{\mathrm{B}}B$, where $\mu_{\mathrm{B}} = e\hbar/2m$ is the Bohr magneton and $g \approx 2$ is the electron g factor. This transition energy can be detected by absorption of the electromagnetic waves propagating parallel to the magnetic field. The resonance absorption condition at $\omega = \omega_{\mathrm{ESR}}$ is $\hbar\omega_{\mathrm{ESR}} = g\mu_{\mathrm{B}}B$.

---

[33]Two spin energy states having the same energy.

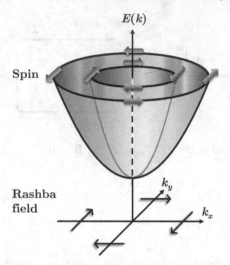

FIGURE 3.9
Energy dispersion of the Rashba Hamiltonian. Fermi surface of two concentric circles of radius $k_\pm$ with the spin structure indicated.

Those two publications, however, prompted Bychkov and Rashba in 1984 to propose a special choice of spin-orbit interaction as the origin of the observed zero magnetic field spin splitting[34], later having some experimental support in their favor[35]. Spin-orbit interaction (SOI) plays a crucial role in the study of electrical manipulation of spin rotation since the orientation of electron spins is coupled with the orbital motion of electrons. As Rashba noted later[36], the spin-orbit coupling that originally played a marginal role in condensed matter physics, went global thanks to those works by Bychkov and Rashba. The reason for this interest is because of the emerging field of semiconductor spintronics (or, spin

[34]Properties of a 2D electron gas with lifted spectral degeneracy, by Yu. A. Bychkov and E.I. Rashba, *JETP Lett.* **39**, 78 (1984); Oscillatory effects and the magnetic susceptibility of carriers in inversion layers, *J. Phys. C: Solid State Phys.* **17**, 6039 (1984).

[35]Zero-magnetic-field spin splittings in $Al_x Ga_{1-x} As/GaAs$ heterojunctions, by P. Ramvall, B. Kowalski, and P. Omling, *Phys. Rev. B* **55**, 7160 (1997).

[36]Spin-orbit coupling goes global, by E.I. Rashba, *J. Phys.: Condens. Matter* **28**, 421004 (2018).

electronics)[37], where this type of spin-orbit interaction plays a crucial role.

The SOI in semiconductor heterostructures can be caused by an electric field perpendicular to the electron plane. Riding on an electron, this electric field will be *felt* as an effective magnetic field lying in the electron plane. The SOI is described by the Hamiltonian[38]

$$\mathcal{H}_{SO} = \alpha \left(\vec{k} \times \vec{\sigma} \right)_z = i\alpha \left(\sigma_y \frac{\partial}{\partial x} - \sigma_x \frac{\partial}{\partial y} \right).$$

Here the z axis is chosen perpendicular to the two-dimensional electron gas (in the xy plane), α is the spin-orbit coupling constant, which is sample dependent and is proportional to the interface electric field that confines the electrons in the xy plane, $\vec{\sigma} = \left(\sigma_x, \sigma_y, \sigma_z \right)$ denotes the Pauli spin matrices, and $\vec{k}$ is the planar wave vector (the momentum of a quantum particle is related to the wave vector as $\vec{p} = \hbar\vec{k}$ or $p = 2\pi\hbar/\lambda$, where λ is the wavelength). The effective magnetic field (the Rashba field) is[39] $\mathbf{B}_{SO} = (2/\hbar)\alpha_{SO}\vec{k} \times \hat{z}$ and depend on the magnitude and direction of the 2D wave vector $\vec{k}$. The corresponding energy dispersion is

$$E^{\pm}(k) = \frac{\hbar^2}{2m} k^2 \pm \alpha k$$

and displayed in Fig. 3.9. The resulting spin splitting is given as $\Delta_{SO} = 2\alpha k$, which is isotropic and linear in k and occurs at $B = 0$. When an electron propagates in a system, the Rashba SO coupling can result in spin precession of the electronic current along its propagating direction due to the interference of two spin-splitting electronic

---

[37]Review on spintronics: Principles and device applications, by A. Hirohata et al., *J. Magnetism and Magnetic Materials*, **509**, 166711 (2020); *Spintronics, Physics in Canada*, edited by T. Chakraborty and K. Hall, vol. 63, No. 2, March/April 2007; Spintronics, by D. Grundler, *Physics World* **15**, 39 (2002); Manipulating spin-orbit interaction in semiconductors, by M. Kohda, T. Bergsten, and J. Nitta, *J. Phys. Soc. Jpn* **77**, 031008 (2008).

[38]This form of SOI can be derived from the nonrelativistic approximation of the Dirac equation and is a standard form that is used in other many-body systems, e.g., Spin-orbit correlations in nuclear matter, by T. Chakraborty and M.L. Ristig, *Lett. Nuovo Cimento Soc. Ital. Fis.* **27**, 65 (1980).

[39]Conduction-subband anisotropic spin splitting in III-V semiconductor heterojunction, by E.A. de Andrada e Silva, *Phys. Rev.* **46**, 1921 (1992).

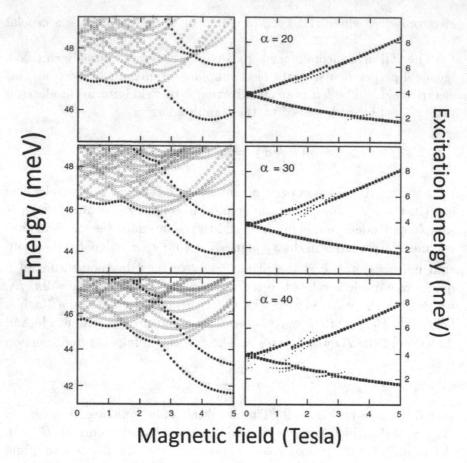

FIGURE 3.10
Energy spectra (left panel) and the optical transition energies (right panel) for four interacting electrons confined in a quantum dot (InAs) for various values of SO coupling strength α. The solid dots in the energy spectrum identify the energy levels involved in transitions that correspond to the lowest branches of the absorption spectra (in the right panel). In the right panels, the size of the points in the figures is proportional to the calculated intensity.

waves[40]. Experimentally, values for α given by Nitta et al.[41] and Heida et al.[42] range between 0.5 and 1×10^{-11} eV m, corresponding to an energy splitting $\Delta_{SO} = 1.5 - 6$ meV. Details about the experimental aspects of the Rashba spin-orbit coupling and its relation to other spin related phenomena can be found in the literature[43].

The effects of Rashba SOI in parabolic quantum dots containing *interacting* electrons were studied in detail by Chakraborty et al.[44]. The energy spectra and the optical transition energies for a quantum dot containing four interacting electrons are shown in Fig. 3.10. The energy spectra are clearly complex with level repulsions and level crossings that are a direct influence of the Rashba SOI. As a consequence, the optical absorption spectra reveal anticrossings and appearance of new modes. Evaluation of these spectra requires diagonalization of 'monster matrices'[45]. Just as the optical transitions in the quantum dot in the absence of SO interaction (Fig. 3.2) revealed interesting facts about the electronic properties, experimental observation of the optical transitions in the present case would also be very exciting. Other novel aspects of quantum dots with the Rashba interaction can be found in the literature[46].

---

[40]Spin orientation and spin precession in inversion-symmetric quasi-two-dimensional electron systems, by R. Winkler, *Phys. Rev. B* **69**, 045317 (2004); Spin precession due to spin-orbit coupling in a two-dimensional electron gas with spin injection via ideal quantum point contact, by M.-H. Liu, C.-R. Chang, and S.-H. Chen, *Phys. Rev. B* **71**, 153305 (2005).

[41]Gate control of spin-orbit interaction in an inverted $In_{0.53}Ga_{0.47}As/In_{0.52}Al_{0.48}As$ heterostructure, by J. Nitta, T. Akazaki, H. Takayanagi, and T. Enoki, *Phys. Rev. Lett.* **78**, 1335 (1997).

[42]Spin-orbit interaction in a two-dimensional electron gas in a InAs/AlSb quantum well with gate-controlled electron density, by J.P. Heida, et al., *Phys. Rev. B* **57**, 11911 (1998).

[43]See for example, New perspectives for Rashba spin-orbit coupling, by A. Manchon, H.C. Koo, J. Nitta and R.A. Duine, *Nature Materials* **14**, 871 (2015).

[44]Optical signatures of spin-orbit interaction effects in a parabolic quantum dot, by T. Chakraborty and P. Pietiläinen, *Phys. Rev. Lett.* **95**, 136603 (2005).

[45]Energy levels and magneto-optical transitions in parabolic quantum dots with spin-orbit coupling, by P. Pietiläinen and T. Chakraborty, *Phys. Rev. B* **73**, 155315 (2006).

[46]Enhanced Rashba effect for hole states in a quantum dot, by A. Manaselyan and T. Chakraborty, *EPL* **88**, 17003 (2009); Spin-orbit interaction induced singlet-triplet resonant Raman transitions in quantum dot helium, by A. Manaselyan, A. Ghazaryan, and T. Chakraborty, *EPL* **99**, 17009 (2012).

3.6   Spin textures and topological charge

Here we briefly describe the spin textures in quantum dots induced by
the spin-orbit interactions (SOI) in 2D nanostructure. We have considered the Rashba SOI as well as the linear Dresselhaus SOI[47]. The corresponding Hamiltonians are: for the Rashba SOC, $H_R = \alpha(\sigma_x p_y - \sigma_y p_x)$,
where the σs are, as usual, the Pauli spin matrices and $p_{x,y}$ are the
components of the planar momentum. For the linear Dresselhaus SOC,
$H_D = \beta(\sigma_y p_y - \sigma_x p_x)$. Here α, β are the coupling strengths of the SOCs.

The Hamiltonian in a QD with SOCs is $\mathcal{H} = \mathcal{H}_0 + \mathcal{H}_C$, where

$$\mathcal{H}_0 = \frac{\mathbf{P}^2}{2m^*} + \frac{m^*}{2}\left(\omega_x^2 x^2 + \omega_y^2 y^2\right) + \frac{\Delta}{2}\sigma_z + \mathcal{H}_R + \mathcal{H}_D \quad (3.1)$$

is the non-interacting Hamiltonian, and $\mathcal{H}_C$ is the Coulomb interaction
term. In Eq. 3.1, Δ is the Zeeman coupling, ω_x and ω_y describe the
parabolic confinements in x and y direction, respectively. The confinement length can be defined as $R_{x,y} = \sqrt{\hbar/(m^*\omega_{x,y})}$. The Coulomb
Hamiltonian, and the lengthy expression for the Coulomb interaction
matrix elements $V_{i,j,k,l}$ can be found in the literature (see Footnote 17).
The full Hamiltonian is diagonalized in the basis of the two-dimensional
quantum oscillator. Any low-energy state can be obtained numerically,
for instance, $|\Psi\rangle = \sum_{\{j\}} d_j \left|(j_1, s_1), (j_2, s_2), \ldots (j_{N_e}, s_{N_e})\right\rangle$, where N_e
is the electron number and d_j is the coefficient of the many-particle (or a
single-particle) basis obtained by the exact diagonalization. In the many-
particle basis $\left|(j_1, s_1), (j_2, s_2), \ldots (j_{N_e}, s_{N_e})\right\rangle$, (j_n, s_n) are the indices for
the n-th electron of the system, where j_n is the index containing the x, y
quantum numbers of the oscillator and s_n is the spin index. The spin
field $\sigma_\mu(\mathbf{r})$ of $|\Psi\rangle$ is

$$\sigma_\mu(\mathbf{r}) = \sum_{k,l,s,s'} \psi_{k,s}^\dagger(\mathbf{r})\, \sigma_\mu \psi_{l,s'}(\mathbf{r})\, \langle\Psi| c_{k,s}^\dagger c_{l,s'} |\Psi\rangle, \quad (3.2)$$

where $\psi_{k,s}(\mathbf{r}) = \psi_{k_x,k_y,s}(\mathbf{r})$ is the wave function of the two-dimensional
oscillator with spin s.

For the quantum-dot hydrogen (i.e., a one-electron quantum dot),
it can be rigorously proven that the winding number of the vector field

---

[47]Spin-orbit coupling effects in zinc blende structures, by G. Dresselhaus, *Phys.
Rev.* **100**, 580 (1955).

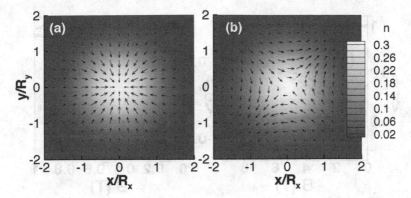

FIGURE 3.11
Single-electron InAs quantum dot of radius $R_x = R_y = 15$ nm, $B = 0.1$ Tesla for (a) Rashba SOC only, $\hbar g_1 = 40$ nm·meV, and (b) for Dresselhaus SOC only, $\hbar g_2 = 20$ nm·meV. Electron density is shown as contours and in-plane spin fields are shown by arrows.

(the in-plane spin field) $(\sigma_x(\mathbf{r}), \sigma_y(\mathbf{r}))$ on an enclosed contour around the origin[48], which is defined as

$$q = \frac{1}{2\pi} \oint \frac{\sigma_x(\mathbf{r})\, d\sigma_y(\mathbf{r}) - \sigma_y(\mathbf{r})\, d\sigma_x(\mathbf{r})}{[\sigma_x(\mathbf{r})]^2 + [\sigma_y(\mathbf{r})]^2}, \tag{3.3}$$

is an integer. More precisely, $q = 1$ if Rashba SOC is present since the in-plane spin rotates clockwise by 2π if we move around the center of the dot in a clockwise direction. On the other hand, $q = -1$ when the Dresselhaus SOC being present, along the same line of reasoning, that the in-plane spin field rotates anticlockwise by 2π if we move around the center in a clockwise direction. The spin textures are displayed in Fig. 3.11.

We thus define the topological charge of the spin field by this winding number. The intrinsic topological property is related to the symmetry of the two SOCs: $[\mathcal{H}_R, L_z + \hbar\sigma_z/2] = 0$ and $[\mathcal{H}_D, L_z - \hbar\sigma_z/2] = 0$. These topological textures of the spin fields are robust against the geometry of the dot. Moreover, when both SOCs are present, we find that $q = -\mathrm{sgn}(g)$ where g is the Landé factor of the material, when the magnetic field approaches infinity.

---

[48]Tuning the topological features of quantum-dot hydrogen and helium by a magnetic field, by W. Luo and T. Chakraborty, *Phys. Rev. B* **100**, 085309 (2019).

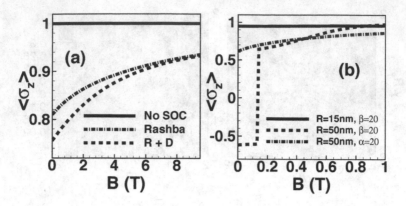

FIGURE 3.12

(a) $\langle \sigma_z \rangle$ in a single-electron InAs dot ($R = R_x = R_y = 15$ nm) without SOC, with Rashba SOC only ($\hbar g_1 = 40$ nm·meV), and with both Rashba and Dresselhaus SOCs ($\hbar g_1 = 40$ nm·meV, $\hbar g_2 = 20$ nm·meV). (b) The sign of $\langle \sigma_z \rangle$ in a large InAs dot are different for different SOCs. Here g_1 and g_2 are given in units of nm·meV/$\hbar$.

The topological charge is closely related to the sign of $\langle \sigma_z \rangle$ in a large dot, allowing a probe of its topological properties[49]. Figure 3.12 demonstrates that the sign of $\langle \sigma_z \rangle$ for a InAs dot of radius $R_x = R_y = 50$ nm in a weak magnetic field $B < 0.14$ Tesla allows one to determine the type of the dominant SOC and thus, indirectly, the topological charge of the dot. We note that for a material with $g > 0$, the reversal of $\langle \sigma_z \rangle$ will instead occur for a dot with dominant Rashba SOC. Information stored as topological charge in a quantum dot system can thus be accessed by measuring the sign of $\langle \sigma_z \rangle$ in a weak magnetic field.

In a quantum-dot helium (i.e., a two-electron dot), the overall winding number can have different properties than those of the single-elctron case (quantum-dot hydrogen). This is because changing the number of electrons affects the winding number due to the Coulomb interaction and the z component of the angular momentum $\langle L_z \rangle$. This is easily understood because $\langle L_z \rangle$ varies at most once in the quantum dot hydrogen, while the Coulomb interaction makes $\langle L_z \rangle$ vary stepwise up to $-\infty$ with increase of the magnetic field. The density profile and the spin texture influence each other in the presence of the Coulomb interaction.

---

[49]Unique Spin Vortices and Topological Charges in Quantum Dots with Spin-orbit Couplings, by W. Luo, et al., *Sci. Rep.* **9**, 672 (2019).

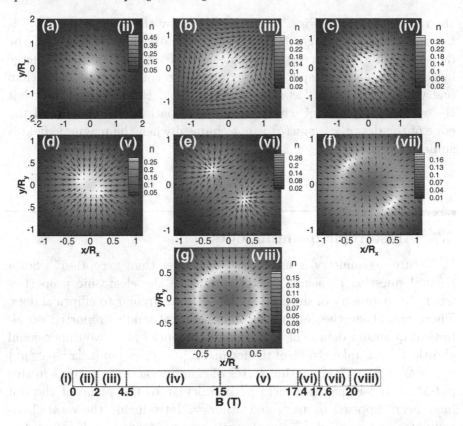

FIGURE 3.13

The density profiles of a two-electron InAs dot, $R_x = R_y = 15$ nm, with SOCs $\hbar g_1 = \hbar g_2 = 20$ nm·meV. The magnetic fields are (a) 1 Tesla; (b) 3.7 Tesla; (c) 6 Tesla; (d) 12 Tesla; (e) 17.5 Tesla; (f) 18 Tesla; (g) 23 Tesla. The Roman numerals correspond to the states indicated in (b) to (g). The state (i) at $B = 0$ Tesla is not shown since it is a degenerate state due to time reversal symmetry.

When $\langle L_z \rangle$ is changed from an integer by the spin-orbit couplings, the rotational symmetry is broken which induces strong density deformation. The two SOCs and the Coulomb interaction jointly influence the spin textures and the density profiles with increasing magnetic field. The density profile evolves with the change of $\langle L_z \rangle$: the density merges to a dot or a ring when $\langle L_z \rangle$ enters a plateau near an integer as shown in Figs. 3.13 (a) and (g). But the dot-shape is stretched [Fig. 3.13 (b)] and the ring-shape splits [Figs. 3.13 (e) and (f)] by the SOCs when $\langle L_z \rangle$ is far away from an integer between two plateaus. The evolution of the density profile with increasing magnetic field thus enters a split-merge cycle. The associated spin textures are more complex and more than one vortex can appear, as shown in Fig. 3.13. However, the rule that the winding number, of which the integral path is chosen around the edge of the dot, $q = -\text{sgn}(g)$ is not changed when the magnetic field is sufficiently strong.

3.7   Anisotropic quantum dots

In Nature, asymmetry is more the rule rather than exception[50]. So, a natural question to ask is what happens to the electronic properties when the symmetry of the dot is reduced from circular to elliptical dots. There have been theoretical and experimental studies reported on elliptical quantum dots. The confinement potential of a two-dimensional elliptical dot is taken to be of the form, $V_{\text{conf}}(x,y) = \frac{1}{2}m\left(\omega_x^2 x^2 + \omega_y^2 y^2\right)$ where $\omega_x \neq \omega_y$. The quantum states of non-interacting electrons in this potential and a magnetic field perpendicular to the plane of the dot have been reported by us[51], and others[52]. Introducing the rotated coordinates and momenta[53], the electron energy is given by $E\left(n_x, n_y\right) = \left(n_x + \frac{1}{2}\right)\hbar\omega_x + \left(n_y + \frac{1}{2}\right)\hbar\omega_y$, i.e., the sum of two independent harmonic oscillators (Fig. 3.14).

As expected, the breaking of circular symmetry in the dot results in the lifting of degeneracies at $B = 0$[54], that are otherwise present in a circular dot. Theoretical studies also indicated that, introduction of anisotropy in a QD results in a major enhancement of the Rashba SO

---

[50] Asymmetry, Symmetry and Beauty, by H. Sabelli and A. Lawandow, *Symmetry* **2**, 1591 (2010).

[51] Electronic properties of anisotropic quantum dots in a magnetic field, by A.V. Madhav and T. Chakraborty, *Phys. Rev. B* **49**, 8163 (1994).

[52] Quantum states of interacting electrons in a 2D elliptical quantum dot, by P.A. Maksym, *Physica B* **249**, 233 (1998).

[53] See Footnote 51.

[54] See Footnote 51, and Optical anisotropy of electronic excitations in elliptical quantum dots, by A. Singha, et al., *Appl. Phys. Lett.* **94**, 073114 (2009).

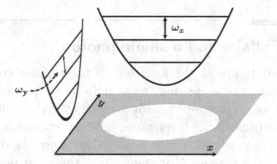

FIGURE 3.14
The energy parabola along the major and minor axes of the anisotropic quantum dot.

coupling effects. This is reflected in the optical absorption spectra of the elliptical dots[55].

Considering the interacting electrons, Maksym[56] noted that only some of the transitions found in a circular dot survive in the lower symmetry of the elliptical dot. In the presence of the Coulomb interaction among the electrons and combined with the Rashba SOI, the eccentricity of the QD causes major modifications of the electron energy spectra. This results in superintense and highly anisotropic optical transitions[57]. The algebra and the computations in these studies are quite involved[58]. Even more so for the interacting electrons with the Rashba SOI included[59]. Interested readers should find the references cited here to be helpful for delving further into the field.

---

[55] Strong enhancement of Rashba spin-orbit coupling with increasing anisotropy in the Fock-Darwin states of a quantum dot, by S. Avetisyan, P. Pietiläinen, and T. Chakraborty, *Phys. Rev. B* **85**, 153301 (2012).

[56] See Footnote 52.

[57] Superintense highly anisotropic optical transitions in anisotropic quantum dots, by S. Avetisyan, P. Pietiläinen, and T. Chakraborty, *Phys. Rev. B* **88**, 205310 (2013).

[58] Fock-Darwin states of anisotropic quantum dots with Rashba spin-orbit coupling, by Siranush Avetisyan, Dissertation, University of Manitoba, 2014.

[59] Magnetization of interacting electrons in anisotropic quantum dots with Rashba spin-orbit interaction, by A. Avetisyan, P. Pietiläinen, and T. Chakraborty, *Physica E* **81**, 334 (2016).

3.8   Secret affairs and a single photon

The quantum dots are at the forefront of the rapidly growing field of quantum information processing, most notably in quantum cryptography. The story of cryptography, as it evolved over the millennia is truly fascinating. Information, be it religious, military, economic, etc. is always important for the powerful, which is why there were always attempts to protect it from the masses. Cryptography – the word derives from the Greek word kryptos meaning hidden, and graphein means writing, has been around almost as long as writing itself. Ciphers have been discovered in Egypt from as early as 2000 BC. In ancient Rome the secret political and military information was regularly encrypted, the most famous being the Caesar cipher (named after Julius Caesar), used since the Gallic wars (58-51 BC). This method of encryption involves replacing each of the letters of the alphabet in the original text by a letter located a set number of places further down the sequence of the letters in the alphabet of the language. For example, with a left shift of four, E would be replaced by A, F would become B, and so on. Of course, the sender and the receiver need to know in advance how many letters to be shifted. Messages are decrypted by doing the reverse substitution.

If in the modern world of cryptography, the most popular person 'Bob' encrypts a message for the other popular character 'Alice' using a Caesar cipher and the message falls into the hands of the eavesdropper 'Eve', she can easily decipher it by simply trying all the 25 possible shifts of the alphabet. Bob could, however, make the message a bit harder by using a more general 'substitution cipher' in which each letter is replaced by another letter (or some symbol). They are however, still not secure because of the fact that, in any given language, different letters are used with different frequencies. So, Eve can make some good guesses and with a few trial and error steps she will be able to crack the code. Despite being insecure, this method of encryption was used widely in Europe until as late as 1800. The famous consequence of the insecurity of substitution ciphers was the beheading of Mary Queen of Scots whose correspondence with plotters against Queen Elizabeth I was deciphered by the code breakers. Edger Allen Poe's mystery story *The Gold Bug* (1843) and A. Conan Doyle's Sherlock Holmes mystery, *Adventure of the Dancing Men* (1903) were based on the decryption of substitution ciphers (Fig. 3.15). For a very enjoyable reading of the history of cryptography

ᛉᚿᚾᛉ ᛏᛉᛏᛉ ᚾᛉ ᚾᛉᚾᛉ ᛏᛉᚱ

FIGURE 3.15
A ciphertext based on *The Adventure of the Dancing Men*, a Sherlock Holmes adventure by Sir Arthur Conan Doyle.

in plain words, *The Code Book* by Simon Singh (Delacorte Press, New York, 2001) is highly recommended.

Today, the ciphers are used for which the algorithms for encrypting and decrypting are known, but one needs a specific set of parameters, the key. The key is shared by what is known as the *key distribution problem*. Interestingly, the key can be distributed publicly but it still remains secure. It can be explained in a simple approach[60] as follows. Alice sends a message by the regular mail, but in order to prevent the mailman from reading the message, she sends it to Bob in a padlocked box. However, the key only belongs to Alice, so when the box arrives to Bob, he puts a second lock with his own key and resends it to Alice who then opens her lock and returns the box to Bob. Then Bob has no difficulty in opening the box and receive the message.

Nowadays, in many public key cryptosystems, the key ingredient is the modular arithmatic[61]. It works as follows: Two integers m, n are considered to be the same if they differ by a multiple of k. We write this as $m \equiv n \pmod{k}$, and say that m and n are *congruent modulo k*. For example, $36 \equiv 10 \pmod{13}$ because $36 - 10$ is an integer multiple of $k = 13$, i.e., $26 = 2 \times 13$. It looks complicated, but unknowingly, we use this method all the time when we use the 12-hour clock: If the time now is 9:00, then 8 hours later it will be 5:00. The clocks actually 'wrap around' every 12 hours. This is arithmetic *modulo 12*. In order to demonstrate how this helps in the public key distribution, we follow the illuminating example from Footnote 60. Alice and Bob agree on the numbers 7 and 11 and evaluate $7^x \pmod{11}$ (which they don't even bother to keep secret). They then execute the following operations:

---

[60]Figuring It Out, by N. Crato (Springer, 2010).
[61]Modular Arithmetic and Cryptography, by J.B. Reade, *The Mathematical Gazette* **72**, 198 (1988).

Alice selects 3 as her secret number	Bob selects 6 as his secret number
Alice calculates $7^3 = 343$	Bob calculates $7^6 = 117649$
$\equiv 2 \pmod{11}$	$\equiv 4 \pmod{11}$
Alice send the result 2 to Bob	Bob sends the result 4 to Alice
Alice uses Bob's results 4 and her secret number 3 and calculates	Bob takes Alice's result 2 and his secret number 6 and calculates
$4^3 = 64 \equiv 9 \pmod{11}$	$2^6 = 64 \equiv 9 \pmod{11}$

They end up with the same number 9 without the need to divulge their personal secret numbers to each other. The result is the number that only Bob and Alice know and will serve as the key for encrypting their messages. All the while Eve is kept in the dark about what is going on!

In recent years, a novel concept of using quantum mechanics for cryptography (quantum cryptography) has become popular[62]. It uses some quantum physical principles to exchange keys between the partners that can be used subsequently to encrypt communication data. The quantum cryptography can therefore be better described as 'Quantum Key Distribution' or QKD for short. QKD uses *single photons* to exchange individual key-bits. A photon is the smallest discrete amount or quantum of electromagnetic radiation. It is the basic unit of all light. The concept of quantized electromagnetic radiation was first introduced by Max Planck in 1900. The term 'photon' was first introduced by G.N. Lewis[63]. The most important feature of QKD is that it is inherently secure. In order to extract the information from a photon one must use the single-photon detector to measure the properties which will destroy the photon. It is also not possible to make an exact replica of this photon. Generation and detection of a single photon is the key here because more than one photon can compromise the security of the communication by allowing Eve to gain information from the extra photons. Detection and emission of a single photon must be an extraordinary feat because the signal will be extremely dim. This can be appreciated from the fact that an ordinary 100 W light bulb will emit approximately 10,000,000,000,000,000,000 (10^{19} or 10 billion billion) photons every second! Semiconductor

---

[62]Key to the quantum industry, by A. Shields and Z. Yuan, *Physics World* **20**, 24 (2007).

[63]The Conservation of Photons, by G.N. Lewis, *Nature* **118**, 874 (1926).

quantum dots play a central role in single-photon detectors and single-photon emitters[64].

3.8.1   Single-photon detectors

An ideal single-photon detector would be the one for which the detection efficiency (the probability that a photon is incident on the detector is successfully detected) is 100%. The detector should also be able to distinguish the number of photons in an incident pulse. Almost all single-photon detectors use the process of converting a photon into an electrical signal. The detector electronics is responsible for ensuring that each photo-generated electrical signal is detected with high efficiency. Single-electron transistors[65] were used to detect single photons in the infrared[66] and far-infrared[67] regions, and in the latter case, with unprecedented sensitivity. The FIR region covers the vibrational spectra of molecules in liquids and gases and therefore the single-photon detector, which also measures the wavelength of the photon, turns out to be an important tool for FIR spectroscopy. An externally applied magnetic field was used to tune the energy so that the cyclotron energy equals the photon energy, the photon is absorbed and an electron is excited into the higher energy level. The single-electron transistor is used to detect that photo-excited electron.

Shields et al.[68] developed a single-photon detector based on a quantum dot arrays to detect individual photons at visible or near-infrared wavelengths (Fig. 3.16). Their device was a GaAs/AlGaAs field-effect transistor containing a layer of InAs QDs separated from a 2DEG in the GaAs channel by a thin AlGaAs barrier. Charge carriers photo-excited by incident light are trapped by the dots which alters the conductivity of the 2DEG and is measured by the device. The conductance is shown to

---

[64]Single-photon sources and detectors, by M.D. Eisaman, et al., *Rev. Sci. Instrum.* **82**, 071101 (2011).

[65]See, for example, Mesoscopic devices, by T.J. Thornton, *Rep. Prog. Phys.* **58**, 311 (1995); The single electron transistor and artificial atoms, by M.A. Kastner, *Ann. Phys.* **9**, 885 (2000).

[66]Nonclassical radiation from a single self-assembled InAs quantum dot, by A.N. Cleland, et al., *Appl. Phys. Lett.* **61**, 2820 (1992).

[67]A single-photon detector in the far-infrared range, by S. Komiyama, et al., *Nature* **403**, 405 (2000).

[68]Detection of single photons using a field-effect transistor gated by a layer of quantum dots, by A.J. Shields et al., *Appl. Phys. Lett.* **76**, 3673 (2000).

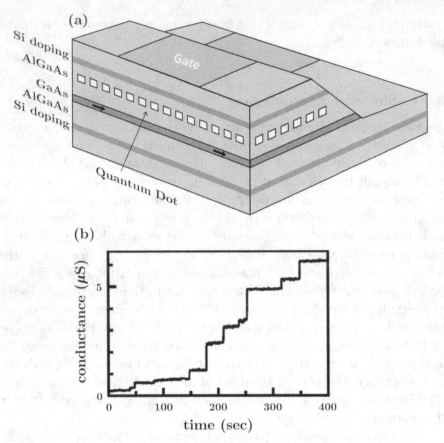

FIGURE 3.16
(a) The quantum-dot FET structure (schematic). Application of a positive bias to the gate charges the underlying dots with electrons. Single
photons liberate a trapped electron, via capture of a photo-excited hole,
resulting in a measurable increase in the conductance of the electron
channel. (b) Time dependence of the FET source-drain conductance under very weak illumination. The conductance displays a series of step-like
rises, each due to capture of a single photo-excited hole by one of the
quantum dots under the gate region.

undergo a series of discrete step-like rises, each due to capture of a single
charge by one of the quantum dots. Shields et al. have also demonstrated
the possibility of detecting the capture or the loss of a single charge in
a quantum dot by sensing a change in the resonant tunneling current

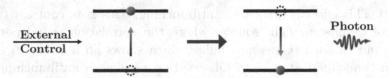

FIGURE 3.17
Single-photon emission process: First the system is excited to a higher energy state, which then relaxes to a lower-energy state by emitting a photon.

through a double-barrier structure[69]. There are a few review articles available in the literature on the work by various groups[70].

3.8.2 Single-photon source

In classical light sources which usually contain a macroscopic number of emitters, the emitted photons obey Poisson statistics. However, creating a light source that will emit *one and only one* photon within a short time interval, on demand, is highly desirable for our purpose. For a secure key distribution in quantum cryptography, the emitter has to provide a pure stream of individual, indistinguishable, photons on demand (or at a high emission rate). There are various systems, such as quantum wells, single molecules, single atoms, and semiconductor quantum dots that are single-photon sources. Although all these single-photon emitters use different material systems, they mostly rely on the same principle of operation. Whenever the single photon is required, some external control sends the system into an excited state that will emit a single photon as the system relaxes to a lower energy state (Fig. 3.17).

In order to demonstrate single-photon generation, various experimental groups have reported photon correlation measurements, in particular, the normalized second-order correlation $g^{(2)}(\tau) = \langle I(t)I(t+\tau)\rangle/\langle I(t)\rangle^2$ where $I(t)$ is the emission intensity at time[71] t. For a truly single-photon source, the correlation between successive photons emitted is zero at

[69]Efficient single photon detection by quantum dot resonant tunneling diodes, by J.C. Blakesley et al., *Phys. Rev. Lett.* **94**, 067401 (2005).

[70]See Footnote 64; and Single-photon detectors for optical quantum information applications, by R. Hadfield, *Nature Photon* **3**, 696 (2009).

[71]Nano-Physics & Bio-Electronics: A New Odyssey, Edited by T. Chakraborty, F. Peeters, and U. Sivan (Elsevier, Amsterdam 2002), Ch. 4.

$\tau = 0$. This effect is known as antibunching. This is in contrast to the case of a coherent light source, where the correlation is independent of τ, and thermal (Gaussian) light, which shows an increase in correlation (bunching) at $\tau = 0$. Observation of photon antibunching has been reported in a wide variety of systems: single atoms[72] (resonance fluorescence of a low-density vapor of Na atoms), a single trapped Mg$^+$ ion[73], single molecules[74], epitaxially grown self-assembled CdSe/Zn (S, Se) quantum dots for temperatures up to 200 K[75], single-photon emission due to single electron-hole pairs photo-excited in a single quantum dot[76], and in self-assembled QDs. A measured value of $g^{(2)}(0)$ as low as 0.05 has been reported[77]. There are several more recent publications including reviews available in the literature for QD single-photon source[78].

In this digital information age, unbeknownst to us we are now all cryptographers! Every time we do online banking, make online shopping and online payments, our computer and the computer of the bank or the store that takes our order, arrange the security by negotiating the encryption using the public key cryptography. For better or worse, quantum cryptography will make it much safer to spend even more money via online!

The field of quantum applications of the photon that includes in addition to cryptography, secure quantum communications over a long range, quantum memory, quantum teleportation through free space, photonic

---

[72]Photon antibunching in resonance fluorescence, by H.J. Kimble, M. Dageneis, and L. Mandel, *Phys. Rev. Lett.* **39**, 691 (1977).

[73]Nonclassical radiation of a single stored ion, by F. Diedrich and H. Walther, *Phys. Rev. Lett.* **58**, 203 (1987).

[74]Photon antibunching in the fluorescence of a single dye molecule trapped in a solid, by Th. Basché, et al., *Phys. Rev. Lett.* **69**, 1516 (1992).

[75]Single-photon emission of CdSe quantum dots at temperatures up to 200 K, by K. Sebald, et al., *Appl. Phys. Lett.* **81**, 2920 (2002).

[76]Triggered single photons from a quantum dot, by C. Santori, et al., *Phys. Rev. Lett.* **86**, 1502 (2001); Single quantum dots emit single photons at a time: Antibunching experiments, by V. Zwiller, et al., *Appl. Phys. Lett.* **78**, 2476 (2001).

[77]Nonclassical radiation from a single self-assembled InAs quantum dot, by C. Becher, et al., *Phys. Rev. B* **63**, 121312(R) (2001).

[78]Engineered quantum dot single-photon sources, by S. Buckley, K. Rivoire, and J. Vuckovic, *Rep. Prog. Phys.* **75**, 126503 (2012); A quantum dot single photon source driven by resonant electrical injection, by M.J. Conterio, et al., *Appl. Phys. Lett.* **103**, 162108 (2013); Electrically driven single-photon source, by Z. Yuan, et al., *Science* **295**, 102 (2002).

quantum networks, a quantum internet, etc. is indeed a very active area of research[79].

3.9 Cascading and burning bright

While the visible region of the electromagnetic spectrum has been studied for centuries, beginning with the early work of Newton, the exploration of the infrared region of the electromagnetic spectrum is still in its infancy. Apparently, there are ambiguities even about the boundaries of this part of the spectral region. The mid-infrared is the spectral region of 3 μm-50 μm, according to the International Organization for standardization. Incidentally, the visible light is in the spectral region of 380-780 nm, while the infrared is in the 780 nm (0.78 μm)-1 mm (1000 μm) range.

3.9.1 Molecular fingerprinting

The mid-IR spectral region is called the molecular-fingerprint region because most molecules display fundamental vibrational absorptions in this region, thus leaving distinctive spectral fingerprints[80]. The mid-IR region also has several windows of transparency (8-13 μm and 3-5 μm) in Earth's atmosphere that allows for environmental gas sensing with extreme accuracy. This is suitable for pollution detection, atmospheric chemistry[81], and a whole host of other applications. Mid-IR lasers are also important for non-invasive medical diagnostics via biomedical spectroscopy[82].

---

[79]The story of light science: From early theories to today's extraordinary applications, by Dennis F. Vanderwerf (Springer, 2017).

[80]Mid-IR Spectroscopic Sensing, by N. Picqué and T.W. Hänsch, *Optics & Photonic News*, June 2019; High Performance Quantum Cascade Lasers, by F. Capasso et al., *Optics & Photonic News*, October 1999.

[81]Methane concentration and isotopic composition measurements with mid-infrared quantum-cascade laser, by A.A. Kosterev, et al., *Opt. Lett.* **24**, 1762 (1999); Sub-part-per-billion detection of nitric oxide in air using a thermoelectrically cooled mid-infrared quantum cascade laser spectrometer, by D.D. Nelson, et al., *Appl. Phys. B* **75**, 343 (2002).

[82]Quantum cascade lasers in biomedical spectroscopy, by A. Schwaighhofer, M. Brandstetter, and B. Lendl, *Chem. Soc. Rev.* **46**, 5903 (2017).

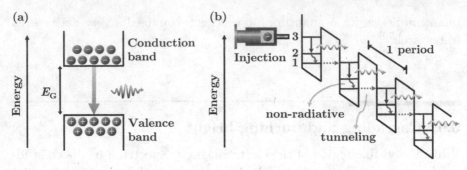

FIGURE 3.18
Schematic diagram of the functioning of a conventional semiconductor laser transition and the quantum cascade transitions. In the standard process a photon is generated from recombination of an electron ($\ominus$) with a hole ($\oplus$). In the QCL, the conduction band profile is engineered to have staircase like structures in an applied voltage. The horizontal line in each quantum well represents the lowest energy subband and the location of the energy level is related to the thickness of the well. In the active regions, each consisting three quantum wells, electrons cascade down the energy levels (transition from the top to the intermediate level) emitting photons in the process and tunnel through the barriers to the next region.

3.9.2 Quantum cascade laser

An important mid-IR laser, the quantum cascade laser (QCL) has made great strides in this technologically important field. The basic principle of the operation of the QCL is based on quantum confinement discussed in Chapter 2. The invention of this laser in 1994 was a great technological achievement[83]. In conventional semiconductor lasers the light originates from recombination of electrons and holes across the energy gap that exists between the conduction band and valence band of the crystal [Fig. QCL (a)]. Each electron can emit only one photon in this process and the wavelength of the light thus originated is fixed by the band gap of the host material.

The QCL, on the other hand, is a special kind of semiconductor laser which exploits the optical transitions in between electronic

---

[83]New frontiers in quantum cascade lasers: high performance room temperature terahertz sources, by M.A. Belkin and F. Capasso, *Phys. Scr.* **90**, 118002 (2015); Quantum cascade lasers: 20 years of challenges, by M.S. Vitiello, et al., *Opt. Express* **23**, 5167 (2015).

subbands. The schematic sketch in Fig. 3.18 (b) shows what happens when an electron is injected in a quantum well in the active region of the laser. In each period of the structure, the electron undergoes a transition between two subbands, subsequently a non-radiative transition to the lowest subband, before tunneling into the upper level of the next quantum well that happens to be aligned to that last level. Due to the marvel of energy-band engineering the injected electron, after it has emitted a laser photon in the active region of the device, is reinjected into the next stage that follows, where it emits another photon, and the process repeats, thereby creating the 'cascade' of transitions. The key features of the QCL lies in its high optical power output and their wide ranging tuning range and room temperature operation. The QCL has become an important laser source in the mid-infrared and also in the terahertz frequency range. The QCLs can now deliver high continuous wave high power output at room temperature, and cover a wide frequency range from 3 to 300 μm by simply varying the material components. QCLs have proven to be the most useful source of terahertz (THz) frequency laser light (the *T-rays*)[84]. Since terahertz radiation can penetrate materials such as fabrics, plastics, and biological tissues, terahertz laser source has found applications in security scanning (e.g., airport screening), hidden explosive detection, etc. There are some excellent review articles available in the literature[85] for readers who are interested to learn more about this unique light source.

3.9.3 QCL in a magnetic field

As the central theme of this book involves the magnetic field effects, it is perhaps appropriate to ask, what happens when one applies a strong magnetic field to the QCL. Theoretical studies of magnetic-field induced luminescence spectra in a QCL were reported by Apalkov and

---

[84]Terahertz quantum-cascade lasers, by B.S. Williams, *Nat. Photonics* **1**, 517 (2007); Thermoelectrically cooled THz quantum cascade laser operating up to 210 K, by L. Bosco et al., *Appl. Phys. Lett.* **115**, 010601 (2019).

[85]Multi-wavelength quantum cascade laser arrays, by P. Rauter and F. Capasso, *Laser Photonics Rev.* **9**, 452 (2015); Quantum cascade lasers: from tool to product, by M. Razeghi, et al., *Opt. Express* **23**, 8462 (2015); High-Power Emission and Single-Mode Operation of Quantum Cascade Lasers for Industrial Applications, by M. Troccoli, *IEEE J. Selected Topics in Quan. Electron.* **21**, 1200207 (2015); Next-generation mid-infrared sources, by D. Jung, et al., *J. Opt.* **19**, 123001 (2017).

Chakraborty[86]. A detailed description of the theoretical studies on this topic can be found in a review article by this author[87]. Application of a tilted magnetic field on a QCL is expected to provide a very interesting effect. An externally applied magnetic field that is tilted from the direction perpendicular to the electron plane has two magnetic field components: the parallel field that makes a shift in energy dispersion in addition to a small diamagnetic shift[88], while the perpendicular component of the magnetic field causes quantization of the subbands. As a result of the shifts of the center of the Landau orbit in wave-vector space, combined intersubband-cyclotron transitions, in addition to the intersubband transitions, are allowed (Fig. 3.19). As the subbands quantize into discrete Landau levels, new luminescence peaks appear which corresponds to these new transitions.

Emission spectra as a function of the tilted field are shown in Fig. 3.20 for three values of $B_\perp$, and for different values of the parallel field for a fixed $B_\perp$. For a small $B_\perp$ [Fig. 3.20 (a)], the emission spectra do not feel the Landau quantization and there is only a single peak that broadens with increasing $B_\parallel$. For higher fields, such as $B_\perp = 5$ Tesla [Fig. 3.20 (b)], there are new features in the emission spectra. For a small parallel field, $B_\parallel = 1$ Tesla, the main transitions are between the states with the same Landau index. Hence the presence of a single peak that corresponds to transitions from the zeroth Landau level (LL) of the second subband to that of the first subband. Increasing the parallel field causes the transitions to higher LLs more intense. Appearance of a shoulder at $B_\parallel = 5$ Tesla corresponds to transitions to the first LL of the first subband. Similarly, the peaks at $B_\parallel = 10$ Tesla and 15 Tesla correspond to transitions from the zeroth LL of the second subband to the higher LL of the first subband. The energy separations between the peaks are equal to the separations between the LL of the first subband. In Fig. 3.20 (c), transitions to a non-zero LL are more intense and we find an interplay between the transitions to zero and to the first LLs

---

[86]Magnetic field induced luminescence spectra in a quantum cascade laser, by V.M. Apalkov and T. Chakraborty, *Appl. Phys. Lett.* **78**, 1973 (2001); Influence of disorder and a parallel magnetic field on a quantum cascade laser, by V.M. Apalkov and T. Chakraborty, *Appl. Phys. Lett.* **78**, 697 (2001).

[87]Quantum cascade transitions in nanostructures, by T. Chakraborty and V.M. Apalkov, *Adv. Phys.* **52**, 455 (2003).

[88]See Footnote 86, and also Influence of a magnetic field on electron subbands in a surface space-charge layer, by W. Beinvogl, A. Kamgar, and J.F. Koch, *Phys. Rev. B* **14**, 4274 (1976).

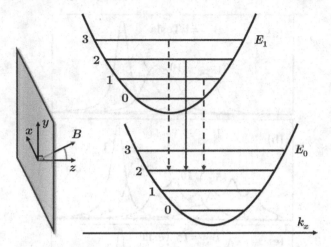

FIGURE 3.19
Schematic illustration of the origin of combined intersubband-cyclotron transitions. Electrons are in the xy-plane while the magnetic field is tilted from the z-axis. The intersubband-cyclotron coupled transitions due to quantization of subbands to the LLs (indicated as 0, 1, 2, ...) are shown here as dashed lines.

with increasing parallel field. For a small $B_\parallel$ of 1 Tesla, there is only a strong transition to the zeroth LL. For $B_\parallel = 5$ Tesla, a small peak that corresponds to transitions to the first LL, appears, and for $B_\parallel = 10$ Tesla and 15 Tesla, transitions to the first LL are the strongest. This leads to the formation of a new narrow peak that corresponds to the transitions to the first LL.

In Fig. 3.21, we show the emission spectra for three values of B_{Tot}, and different values of the tilt angle at a fixed B_{Tot}. For a small field [Fig. 3.21 (a)], there is a red shift of the spectra with increasing parallel field (i.e., increasing tilt angle). For $\theta = 45°$ there is a weak structure that results from the Landau quantization. At higher fields [Fig.3.21 (b), and (c)], we observe the evolution of the spectra from a broad peak at a large angle $\theta = 80°$ (a large parallel field and a small perpendicular field) to a single narrow peak for a small angle. In the latter case, the parallel field is small and all optical transitions happen between the LLs with the same index. For an intermediate tilt angle, there are two peaks that correspond to transitions from the zeroth LL of the second subband to the zeroth and the first LLs of the first subband. The intensity of the

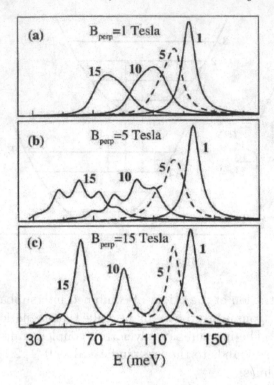

FIGURE 3.20

Luminescence spectra at various values of $B_{\parallel}$ (numbers by the curves in tesla) for a fixed value of $B_{\perp}$.

transitions to the first LL increases with increasing angle, which means an increase of the parallel field. It has its maximum at $\theta = 45°$ and for a total field of 21 Tesla we see the formation of a strong narrow peak at $\theta = 45°$ associated with the suppression of the original peak corresponding to the transition to the zeroth LL of the first subband.

These results led us to conclude that by suitably tuning the externally applied tilted field, optical transitions due to coupled intersubband-cyclotron transitions that is as strong as the zero field case (but at different energies) can be achieved.

Experimentally, Landau quantization in a strong magnetic field was found to enhance the terahertz intersubband luminescence in a quantum cascade structure[89]. The laser emission intensity increases substantially for the non-zero magnetic field.

---

[89]Magnetic-field-enhanced quantum-cascade emission, by J. Ulrich, et al., *Appl.*

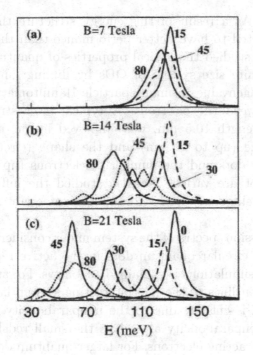

FIGURE 3.21
Luminescence spectra at various values of the tilt angle (numbers by the curves) for total fields of (a) B = 7 Tesla, (b) B = 14 Tesla and (c) B= 21 Tesla.

3.9.4 QCL with quantum dots

Finally, quantum dots replacing the quantum wells in a QCL could be a promising mid-infrared source. They were theoretically proposed by Apalkov and Chakraborty[90] and a few other authors[91]. In recent years there has been considerable progress in quantum-dot laser research. This is because of the discrete atom-like structure of energy levels in

Phys. Lett. **76**, 19 (2000); Terahertz quantum cascade lasers in a magnetic field, by V. Tamosiunas, et al., *Appl. Phys. Lett.* **83**, 3873 (2003).

[90]Luminescence spectra of a quantum-dot cascade laser, by V.M. Apalkov and T. Chakraborty, *Appl. Phys. Lett.* **78**, 1820 (2001); Optical properties of a quantum-dot cascade structure, by V.M. Apalkov and T. Chakraborty, *Physica E* **14**, 294 (2002).

[91]Novel quantum box intersubband lasing mchanism based on image charges, by S.J. Lee and J.B. Khurgin, *Appl. Phys. Lett.* **69**, 1038 (1996); Quantum-dot cascade laser: Proposal for an ultralow-threshold semiconductor laser, by N.S. Wingreen and C.A. Stafford, *IEEE J. Quant. Eletron.* **33**, 1170 (1997); Quantum dot cascade laser: Arguments in favor, by I.A. Dimitriev and R.A. Suris, *Physica E* **40**, 2007 (2008).

quantum dots. As a result of this energy structure the quantum-dot lasers are expected to have better performance than the quantum-well lasers. We have studied the optical properties of quantum dots cascade system for a finite size system of QDs by finding numerically eigenfunctions and eigenvalues of many-particle Hamiltonian with Coulomb interaction between the electrons. A typical laser structure[92], emitting at a wavelength 10.5 μm, was employed in our numerical simulations. The size (up to 20 nm) and the shape (circular or elliptic) of the quantum dots and the number of electrons (up to 9 electrons) in an active dot are varied. We also studied the influence of external magnetic field on the emission spectra of quantum dot cascade structures.

Typical emission spectra of the system under consideration are shown in Fig. 3.22 for circular quantum dots in the active region. The main outcome of our simulations is explained as follows. For smaller quantum dots the emission lines of the non-interacting system have the internal structure which is entirely due to the nonparabolicity of electron dispersion. The nonparabolicity also gives the small redshift of emission line for non-interacting electrons. For larger quantum dots there is only a single line for all number of electrons of the non-interacting system. Such a structure of the emission line is because the non-parabolicity in this case becomes less important due to smaller values of quantum dot confinement energies. The electron-electron interactions result in a huge blueshift of the emission spectra compared to the results for noninteracting electrons. The blueshift becomes smaller for larger quantum dots and also decreases for elliptic dots. This is due to weaker interactions between the electrons when the in-plane spreading of the electron wave function becomes larger. The interactions between the electrons also modify the shape of the emission line. This becomes more pronounced for smaller quantum dots. For larger quantum dots the disorder makes the emission line almost single peaked, especially for larger number of electrons when the interaction between the electrons have less affect on the properties of the quantum dot system[93]. For smaller quantum dots, the change of the shape of the dots changes the shape of the emission line considerably. However, for larger dots the shape of the emission line is less sensitive to

---

[92]Long-wavelength ($\lambda \sim 10.5\,\mu$m) quantum cascade lasers based on a photon-assisted tunneling transition in strong magnetic field, by S. Blaser, L. Diehl, M. Beck, and J. Faist, *Physica E* **7**, 33 (2000).

[93]See Footnote 90.

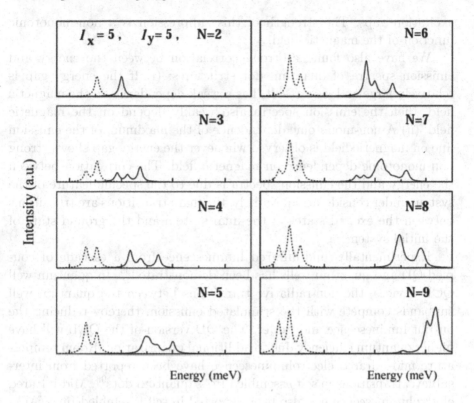

FIGURE 3.22
Luminescence spectra of a quantum cascade laser with circular (size $l_w = l_y = 5$ nm) quantum dots containing $N = 2-9$ electrons in the active region. Solid and dotted lines correspond to interacting and non-interacting electron systems respectively.

the shape of the quantum dots and the line has almost the single peak for both circular and elliptic dots.

The magnetic field effects on intersubband transitions in quantum dots embedded in a quantum cascade structure subjected to a strong magnetic field applied perpendicular to the quantum dot plane was also investigated by Chakraborty and Apalkov[94]. One of the important observations which followed from our simulation was that at a given value of the magnetic field the inter-electron interactions tend to suppress the

---

[94]Magnetic field effects on intersubband transitions in a quantum nanostructure, by T. Chakraborty and V.M. Apalkov, *Physica E* **16**, 253 (2003).

excitation gaps. The strength of this suppression is a non-monotonic function of the magnetic field.

We have also found a strong correlation between the energy and emission spectra of quantum dot structures: (i) If the energy gap is relatively large and, as a result, has a weak dependence on the magnetic field, then the emission spectra also weakly depend on the magnetic field. (ii) A non-monotonic dependence of the maximum of the emission line on magnetic field is observed whenever the energy gap shows strong non-monotonic dependence on magnetic field. The correlation between the energy and the emission spectra is due to the specific feature of the system under consideration. Namely, the main transitions are transitions between the excited states of the final system and the ground state of the initial system.

Experimentally, mid-infrared luminescence from a cascade of coupled QD and quantum wells has been demonstrated[95]. In quantum well QCL devices, the non-radiative transitions between the quantum well subbands compete with the stimulated emission, thereby reducing the output luminescence and power. The QD version of the QCL will have higher quantum efficiency due to additional quantization. Room temperature mid-infrared electroluminescence have been reported from inter-sublevel transitions in self-assembled InAs quantum dots[96]. Mid-infrared electroluminescence has also been reported in self-assembled InAs/AlAs quantum dots by several other groups[97]. Much more progress is expected in the coming years.

---

[95]Midinfrared intraband electroluminescence from AlInAs quantum dots, by N. Ulbrich, et al., *Appl. Phys. Lett.* **83**, 1530 (2003); Influence of dimensionality on the emission spectra of nanostructures, by V.M. Apalkov and T. Chakraborty, *Appl. Phys. Lett.* **83**, 3671 (2003); Temperature-induced broadening of the emission lines from a quantum-dot nanostructure, by V.M. Apalkov, T. Chakraborty, N. Ulbrich, D. Schuh, J. Bauer, and G. Abstreiter, *Physica E* **24**, 272 (2004).

[96]Room temperature midinfrared electroluminescence from InAs quantum dots, by D. Wasserman, et al., *Appl. Phys. Lett.* **94**, 061101 (2009).

[97]Multiple wavelength anisotropically polarized mid-infrared emission from InAs quantum dots, by D. Wasserman, C. Gmachl, S.A. Lyon, and E.A. Shaner, *Appl. Phys. Lett.* **88**, 191118 (2006); Electroluminescence of a quantum dot cascade structure, by S. Anders, et al., *Appl. Phys. Lett.* **82**, 3862 (2003).

4

Quantum rings: Dynamic unity of polar opposites

In the book, 'The Tao of Physics: An Exploration of the Parallels between Modern Physics and Eastern Mysticism', the author Fritjof Capra observed, 'Suppose you have a ball going around a circle. If this movement is projected onto a screen, it becomes an oscillation between two extreme points. ... In any projection of that kind, the circular movement will appear as an oscillation between two opposite points, but in the movement itself the opposites are unified and transcended.'[1]. Surely, an apt description of motion in a ring.

A small and perfect metal ring (or a cylinder) in the absence of disorder, when placed in a perpendicular magnetic field at very low temperature, will exhibit oscillatory behavior of the equilibrium properties such as the average energy or magnetization as a function of the field. That was the conclusion from a detailed theoretical study by Dingle[2] in 1952. The 'smallness' was defined as the case where the radius of the system is much smaller than the electronic orbits (that happens at low magnetic fields).

Thus began the quest to uncover the elusive behavior of mini rings in a magnetic field, that we are about to discuss in this chapter. We begin with the ideal, strictly one-dimensional metal rings.

---

[1]The Tao of Physics: An Exploration of the Parallels Between Modern Physics and Eastern Mysticism by Fritjof Capra (Shambhala, 1976); Eastern mysticism and the alleged parallels with physics, by E.R. Scerri, Amer. J. Phys. **57**, 687 (1989).

[2]Some magnetic properties of metals IV. Properties of small systems of electrons, by R.B. Dingle, Proc. Roy. Soc. A, **212**, 47 (1952).

DOI: 10.1201/9781003090908-4

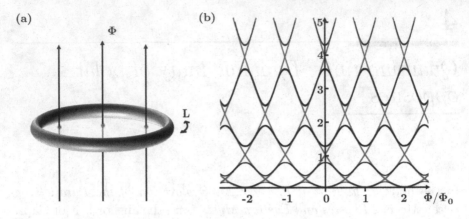

FIGURE 4.1
(a) One-dimensional ring threaded by the magnetic flux Φ. (b) Energy levels (arbitrary units) as function of the magnetic flux. Light and dark lines correspond to a clean and a disordered ring, respectively.

4.1 Tireless electron running around in circles

An important work on this topic was by Büttiker, Imry and Landauer[3] in 1983, who introduced the concept of the 'persistent current' in a strictly one-dimensional metal ring. As shown in Fig. 4.1 (a), in a conducting structure that has the topology of a ring, threaded by a magnetic flux Φ (total magnetic field passing through the area of the ring), the electron motion is equivalent to the motion in an infinite system with a periodic potential. Then the energies, and hence all equilibrium properties of the ring are periodic in Φ with period Φ_0, the flux quantum [Fig. 4.1 (b)]. In this situation there exists a persistent current $I = -\partial E/\partial \Phi$, where E is the ground state energy (see Box 4.1). This current will persist in time since the system is already in the *lowest* energy state available. The persistent current is a thermodynamic equilibrium property and a direct consequences of quantum coherence over a length scale given by the ring's circumference L. It is constant in time and not driven by a power source that is external to the ring (other than that the external flux is

---

[3] Josephson behavior in small normal one-dimensional rings, by M. Büttiker, Y. Imry, and R. Landauer, *Phys. Lett.* **96 A**, 365 (1983).

on). It can be probed by measuring the miniscule magnetic moment. A direct probe of the current is not possible, since the current flows only around a closed loop and any probe on the loop will destroy the effect. The effect of temperature on the current is also quite significant. Therefore, the experiments have to be performed at very low temperatures. A more in-depth historical account of the mesoscopic rings can be found in a recent book on this subject[4].

Box 4.1 Persistent current in an ideal ring:

We assume that the magnetic flux Φ (total field that passes through a surface area) threads the ring axially but the electrons move in a field-free region. In that case, the electron motion in the ring is equivalent to that in an infinite system with a periodic potential. The presence of the magnetic field can be gauged away such that the field does not appear explicitly in the calculations, but appears instead via the flux-modified boundary conditions:

$$\psi(L) = \exp\left[\frac{i2\pi\Phi}{\Phi_0}\right]\psi(0); \quad \frac{d\psi}{dx}\bigg|_{x=L} = \exp\left[\frac{i2\pi\Phi}{\Phi_0}\right]\frac{d\psi}{dx}\bigg|_{x=0},$$

where $\Phi_0 = h/|e|$ is the flux quantum, x is the coordinate along the circumference, and L is the circumference of the ring. These equations imply that all equilibrium properties of the ring are periodic in Φ with a period Φ_0. A non-integer flux is therefore mathematically equivalent to a change in boundary conditions of the system.

We can identify $2\pi\Phi/\Phi_0$ with the momentum[5] in the one-dimensional Bloch problem. The energy levels of the ring form microbands as a function of Φ with period Φ_0, just as the Bloch electron bands in the extended k-zone [Fig. 4.1 (b)]. The current carried by level E_n at $T = 0$ is, $I_n = -ev_n/L, v_n = \frac{1}{\hbar}\frac{\partial E_n}{\partial k_n}$ or, from the analogy above, $I_n = -\partial E_n/\partial \Phi$.

For a free-electron model in a single ring without any disorder,

---

[4] *Physics of Quantum Rings,* edited by V. Fomin (Springer, 2018).

the energy and the corresponding current are

$$E_n = \frac{\hbar^2}{2m} \left[\frac{2\pi}{L} \left(n + \Phi/\Phi_0 \right) \right]^2 \; ; \; I_n = -\frac{2\pi e\hbar}{mL^2} \left[n + \Phi/\Phi_0 \right],$$

where m is the electron mass and $n = 0, \pm 1, \pm 2, ...$. The total current at $T = 0$ is obtained by summing all contributions from levels with energies less than the Fermi energy.

How does one observe the persistent current in this quantum ouroboros[6]? Early experiments to detect the persistent current were carried out on relatively large (μm size) metallic rings containing a large number of electrons and impurities[7]. The observed results were however, not in agreement with the theoretical expectations[8] as described above. However, a recent work on metallic (Au) rings[9] has shown a much better agreement with theoretical predictions. A semiconductor ring in a GaAs/GaAlAs heterojunction[10] (also of μm size but in the *ballistic* regime) displayed the persistent current to be periodic with a period Φ_0 and the amplitude $0.8 \pm 0.4 \, ev_F/L$, in agreement with the theoretical predictions. These experiments with their tantalizingly close to theoretical results have inspired a large number of researchers to report a variety of theoretical studies involving various averaging procedures, dependence on the chemical potential, temperature, different realizations of disorder, and often conflicting conclusions about the role of electron-electron interactions.

[5]Persistent currents in small one-dimensional metal rings, by Ho-Fai Cheung, et al., *Phys. Rev.* B **37**, 6050 (1988).

[6]The subject of August Kekulé's dream that ostesibly led to his discovery of the benzene ring structure.

[7]Magnetization of mesoscopic copper rings: Evidence for persistent currents, by L.P. Levy et al., *Phys. Rev. Lett.* **64**, 2074 (1990); Magnetic response of a single, isolated gold loop, by V. Chandrasekhar, et al., *Phys. Rev. Lett.* **67**, 3578 (1991).

[8]Normal persistent currents, by U. Eckern and P. Schwab, *Adv. Phys.* **44**, 387 (1995); Persistent currents in mesoscopic metallic rings: Ensemble average, by G. Montambaux, H. Bouchiat, et al., *Phys. Rev.* B **42**, 7647 (1990).

[9]Persistent current in normal metal rings, by H. Bluhm, et al., *Phys. Rev. Lett.* **102**, 136802 (2009).

[10]Experimental observation of persistent currents in GaAs-AlGaAs single loop, by D. Mailly, C. Chapelier, and A. Benoit, *Phys. Rev. Lett.* **70**, 2020 (1993).

In order to clearly understand the role of electron-electron interaction in a quantum ring (QR) without getting embroiled by a variety of issues mentioned above, we constructed a model of a QR, inspired by our prior work on quantum dots, that is disorder free, contains only a few interacting electrons and most importantly, can be solved *exactly* (albeit numerically) by the exact diagonalization scheme[11], developed by us earlier in 1992[12] and later reported in more details in 1994[13]. One major advantage of this model is that, as explained below, once the energy levels are thus evaluated, other physical quantities in addition to the persistent current, such as optical absorption, electron spin effects, etc. can also be studied very accurately Over the years, our model has garnered a huge following[14].

4.2 Interacting electrons in a few-electron quantum ring

Just as we treated the quantum dots in the previous chapter as electrons confined in a parabolic potential, we consider here the electrons confined in a potential of the form $v(r) = \frac{1}{2}m\omega_0^2 (r - R)^2$ (Fig. 4.2) to model the quantum ring, where again ω_0 is the confinement potential strength, and R is the radius of the ring. We introduce the parameter $\alpha = \omega_0 m \mathcal{A}/h$, where $\mathcal{A} = \pi R^2$ is the area of the ring. The parameter α is inversely proportional to the width of the ring: large α corresponds to a narrow

---

[11]Generation of Coulomb matrix elements for the 2D quantum harmonic oscillator, by M. Pons Viver and A. Puente, *J. Math. Phys.* **60**, 081905 (2019).

[12]Electronic properties of quantum dots and quantum rings in magnetic fields, by T. Chakraborty and P. Pietiläinen, in *Transport Phenomena in Mesoscopic Systems*, edited by H. Fukuyama and T. Ando (Springer, Heidelberg 1992), p. 61-72.

[13]Electron-electron interaction and the persistent current in a quantum ring, by T. Chakraborty and P. Pietiläinen, *Phys. Rev. B* **50**, 8460 (1994).

[14]See for example, Quantum rings for beginners: energy spectra and persistent currents, by S. Viefers, et al., *Physica E* **21**, 1 (2004); Ground state and far-infrared absorption of two-electron rings in a magnetic field, by A. Puente and L. Serra, *Phys. Rev. B* **63**, 125334 (2001); Collective oscillations in quantum rings: A broken symmetry case, by M. Valin-Rodriguez, A. Puente, and L.I. Serra, *Eur. Phys. D* **12**, 493 (2000); Electronic states of InAs/GaAs quantum ring, by Shu-Shen Li and Jian-Bai Xia, *J. Appl. Phys.* **89**, 3434 (2001); Quantum size effects on the energy levels and far-infrared spectra: Changes from quantum dots to quantum rings, by H. Pan, Int. *J. Mod. Phys. B* **21**, 4715 (2007); Quantum rings under magnetic fields: electronic and optical properties, by Z. Barticevic, M. Pacheco, and A. Latge, *Phys. Rev. B* **62**, 6963 (2000); Single-particle electronic spectra of quantum rings: A comparative study, by J. Simonin et al., *Phys. Rev. B* **70**, 205305 (2004).

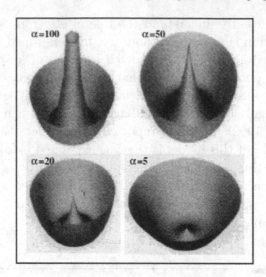

FIGURE 4.2
Confinement potential $\alpha \left(r - R \right)^2$ for different values of α.

path for electrons to traverse, and hence the electron motion is close
to that of a strictly one-dimensional ring discussed above. For small α,
the electron motion is almost two-dimensional. Therefore, our model for
quantum ring can reach the two important limits simply by varying the
parameter α.

The single-electron energy levels are obtained by numerically solving
the QR model and are shown in Fig. 4.3. For $\alpha = 20$, the lower set of
energy levels shown in Fig. 4.3 (b) are similar to those of the ideally
narrow ring, whose energies are, as explained above, given by a set of
translated parabolas [as in Fig.4.3 (a)]. The upper Landau band remains
at higher energies. As α decreases, i.e., the ring becomes wider, the *saw-
tooth* behavior of the narrow ring is gradually replaced by the formation
of Fock-Darwin levels, as in quantum dots. The level crossings are also
quite prevalent in the case of $\alpha = 5$, as shown in Fig. 4.3.

Results for the magnetization $(\mathcal{M})$ and susceptibility (χ) for a three-
and four-electron system (non-interacting) are shown in Fig. 4.4 at various values of the temperature $[k_B T = 0.1$ (dotted line) and 0.2-1.0
(solid lines)] for $\alpha = 20$, when the system is still close to the ideal ring.

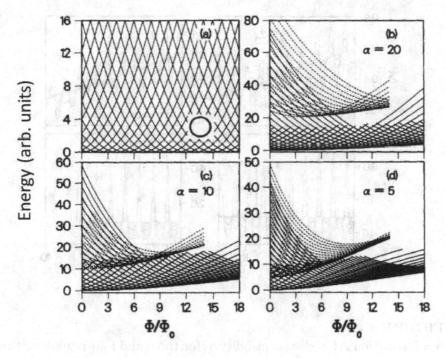

FIGURE 4.3

Energy levels of a single electron versus the magnetic field for (a) ideally narrow ring, and (b)-(d) parabolic confinement model for various of the confinement potential strength. The second Fock-Darwin level is plotted as dotted lines.

Both $\mathcal{M}$ and χ exhibit periodic oscillations (see footnotes 12, 13), as predicted by Dingle, as a function of Φ (albeit with decreasing amplitude as the magnetic field increases. For lower values of α, these oscillations are damped with increasing magnetic field, thus reflecting the behavior in the energy spectrum at those values of α.

Energies of a quantum ring containing four interacting and non-interacting electrons are shown in Fig. 4.5. The electrons are considered to be spinless, and in that case, the only discernible effect of electron-electron interaction is an upward shift of the total energy. This is because in a narrow ring, all close-lying states are in the lowest Landau level and cannot be coupled by the interaction because of the conservation of the

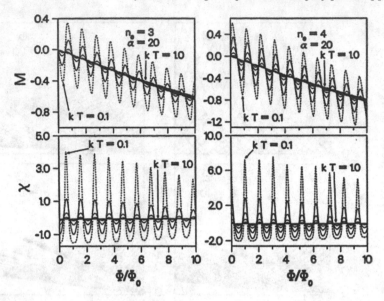

FIGURE 4.4

Magnetization $\mathcal{M}$ and susceptibility χ for three and four noninteracting electrons in a parabolic quantum ring with $\alpha = 20$ for various temperatures.

angular momentum. The excited states seem to shift from the ground state more than that of the corresponding noninteracting case.

Our studies indicated that the electron-electron interaction has no influence on the magnetization which remains periodic with period Φ_0. The technical details of the QR model described here can be found in our 1994 paper (see Footnote 13).

4.3   Optical spectroscopy

Far-infrared (FIR) spectroscopy of μm-size quantum ring arrays that were created in GaAs/AlGaAs heterojunctions was first reported in 1993[15]. The rings were of two different sizes: The outer diameter of the

---

[15]Magnetoplasma resonances in two-dimensional electron rings, by C. Dahl, et al., *Phys. Rev. B* **48**, 15480(R) (1993).

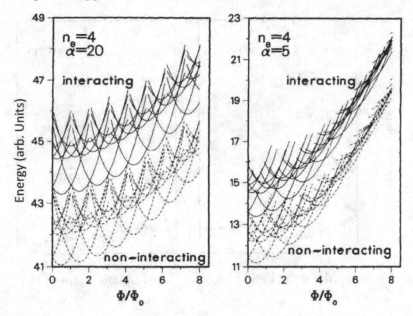

FIGURE 4.5
Energy spectrum of quantum rings containing four non-interacting and interacting spinless electrons for two different widths of the ring.

rings were $\approx 50\,\mu$m in both cases, but the inner diameters were $12\,\mu$m ('broad rings') and $30\,\mu$m ('narrow rings'). The authors described the ring as 'disks with a repulsive scatterer at its center'. These rings contain a large number (of the order of one million) of electrons. The ring patterns are etched into the heterostructure where the corresponding two-dimensional electron gas has the density of 2.3×10^{11} cm^{-2}. The rings are arranged in a square lattice covering an area of several mm^2 in order to increase the signal strength. The observed resonant frequencies are shown in Fig. 4.6. The magnetic field dispersion of the resonant frequencies are significantly different for two different ring geometries.

Halonen et al.[16] developed the theoretical foundations for a quantum ring containing an impurity (modelled by a Gaussian potential) containing upto three (interacting) electrons. The ring is subjected to a

---

[16]Optical-absorption spectra of quantum dots and rings with a repulsive scattering centre, by V. Halonen, P. Pietiläinen, and T. Chakraborty, *Europhys. Lett.* **33**, 377 (1996).

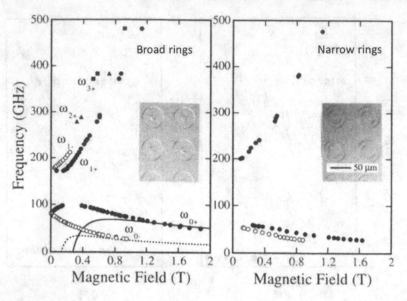

FIGURE 4.6
Resonant frequencies versus the magnetic field (in Tesla) for 'broad' and 'narrow' rings.

perpendicular magnetic field. The absorption energies and intensities of such a system is shown in Fig. 4.7.

The two upper modes of the one-electron spectrum behave almost similar to the experimental results described above. The two lower modes show a periodic structure similar to the case of impurity-free parabolic ring described below. When the number of electrons in the system is increased, the periodic structure of the two lowest modes start to disappear due to electron- electron interaction. In our model, we took the electron spin into consideration, the difference between the one- and two-electron results in Fig. 4.7 is entirely due to the Coulomb force. The lowest mode (which is also the strongest) behaves (even for only three electrons) much the same way as does the lowest mode in the experiment (Fig. 4.6), where the system consists of the order of one million electrons! Interestingly, some of the modes predicted in this model have been confirmed in more recent experiments involving only a few electrons. That will be discussed in the following text.

In a narrow ring without any impurity, the absorption from the ground state can occur with equal probability to the first two excited

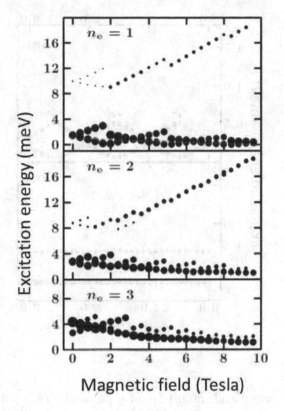

FIGURE 4.7

Absorption energies and intensities of a quantum dot with a Gaussian repulsive scatterer versus the magnetic field. The dot contains one to three electrons. The areas of the filled circles are proportional to the calculated intensity.

states. The presence of impurity will mix the eigenstates of the pure system into new states between which the transitions are allowed. As shown in Fig. 4.8 (a), for an impurity of moderate strength, an appreciable part of the transition probability still goes to the first two excited states. For a strong impurity, absorptions taking to the lowest excited states are more favorable [Fig. 4.8 (b)]. These authors also noticed an interesting feature in this study: in a system with broken rotational symmetry, the transition probability depends strongly on the polarization of the incident light. As an example, if instead of the unpolarized light considered here, we were to consider the case of light polarized along the diameter

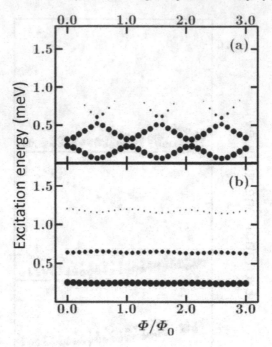

FIGURE 4.8
Absorption energies and intensities of a parabolic ring without any impurity. The two cases are for (a) a weak impurity and (b) for a strong impurity. The area of the filled circles are proportional to the calculated intensity.

passing through the impurity[17], the absorption would occur in the *second* excited state. The other interesting feature observed in Fig. 4.8 was the periodic behavior of absorption energies as a function of the applied magnetic field that follows closely the behavior of the persistent current. A strong impurity flattens the curves of absorption energies that corresponds to blocking the persistent current.

In order to investigate the role of impurity potentials and electron correlations, we consider the ring with four spinless electrons. The effect

---

[17]Electron correlations in quantum ring and dot systems, by P. Pietiläinen, V. Halonen, and T. Chakraborty, in Proc. of the International Workshop on Novel Physics in Low-Dimensional Systems, edited by T. Chakraborty, *Physica B* **212**, 201-327 (1995), p. 256.

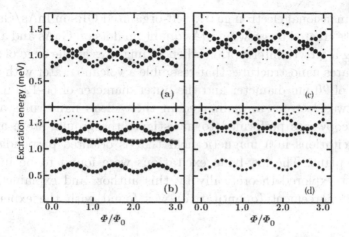

FIGURE 4.9

Absorption energies and intensities four non-interacting [(a) and (b)]
and interacting [(c) and (d)] electrons in a QR versus the magnetic field.
The impurity strength is the same as in the previous figure.

of impurity, in general, can be qualitatively explained by the single-
electron properties. As an example, when we compare Fig. 4.9 (a) and
(b), we notice that lifting of the degeneracy in the energy spectra of non-
interacting electrons [Fig. 4.1 (a)] is reflected by a smoother behavior as
a function of the magnetic field. The sole effect of Coulomb interaction
on the energy spectrum is to shift it upwards and to increase the gap between the ground state and the excited states. As a result, the Coulomb
interaction moves the absorption to higher frequencies [Fig. 4.9 (c) and
(d)]. The effect of electron-electron interaction is best reflected in the
intensity (size of the dots in the figures): for the non-interacting system
[Fig. 4.9 (a) and (b)], the intensity of each absorption mode does not
depend on the magnetic field, but for the interacting system [Fig. 4.9 (c)
and (d)] there is a strong variation of intensity as a function of the field.
The optical-absorption spectra in a quantum ring not only reflects the
behavior of the persistent current, but it also reveals the subtle effects
of the interaction and broken symmetry caused by an impurity.

Quantum rings of nanometer dimensions are usually created as self-
organized[18], or fabricated on GaAs/AlGaAs heterojunctions containing

[18]Spectroscopy of Nanoscopic Semiconductor Rings, by A. Lorke et al., *Phys.
Rev. Lett.* **84**, 2223 (2000); Self-organized InGaAs quantum rings – Fabrication and
spectroscopy, by A. Lorke, et al., *Adv. in Solid State Phys.* **43**, 125 (2003).

a two-dimensional electron gas[19]. Self-organized QRs in InAs/GaAs systems were created by growing InAs quantum dots on GaAs and a process involving a growth interruption when In migrates at the edge of the dot. This creates nanostructures that resemble a volcano crater with the center hole of 20 nm diameter and the outer diameter of 60-120 nm. Only one or two electrons are admitted in these clean nano rings and FIR spectroscopy was performed to investigate the ground state and low-lying excitations in a magnetic field that is oriented perpendicular to the QR plane. The low-lying excitations were found to be unique to QRs first explored theoretically by this author, and explained above. Excellent agreement (quantitative) was found with the experimental results.

4.4 Role of electron spin

At low magnetic fields, electron spins are expected to influence the energy spectrum quite significantly. In Fig. 4.10, we show the ground state energies calculated for up to ten non-interacting electrons in a ring[20]. Clearly, the major consequence of the spin degree of freedom is period and amplitude halving of the energy with increasing number of the flux quanta. It is also strongly particle number dependent. This result for a non-interacting system was reported earlier[21], and can be easily accounted for by simply counting the number of spins, following the Pauli principle, and noting that up and down spins contribute equally to the energy. The particle number (*modulo* 4) dependence can also be trivially explained in this way.

The Coulomb interaction in our model was found to have a profound effect on the energy spectra when the spin degree of freedom was included. The low-lying energy states evaluated for two non-interacting (a) and interacting (b) electron system are shown in Fig. 4.11. As the

---

[19] Kondo effect in a few-electron quantum ring, by U.F. Keyser, et al., *Phys. Rev. Lett.* **90**, 196601 (2003).

[20] Fractional oscillations of electronic states in a quantum ring, by K. Niemelä, et al., *Europhys. Lett.* **36**, 533 (1996).

[21] Period and amplitude halving in mesoscopic eings with spin, by D. Loss and P. Goldbart, *Phys. Rev. B* **43**, 13762 (1991).

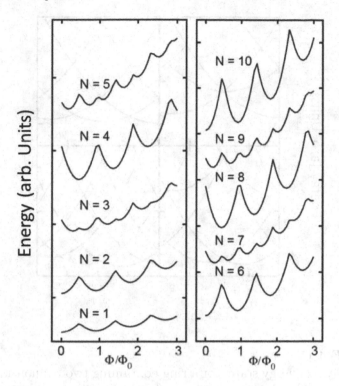

FIGURE 4.10
Ground state energy as a function of Φ/Φ_0 for up to ten non-interacting electrons with electron spin effect included. The energies are scaled to illustrate the dependence of periodicity on the number of electrons in the ring.

flux is increased, the angular momentum quantum number (L) of the ground state of a non-interacting system is increased by two, i.e., the ground state changes as 0, 2, 4, ... The period is, as usual, one flux quantum. The ground state of the non-interacting system is always a spin-singlet. The first excited state is spin degenerate.

As the Coulomb interaction is turned on, the singlet-triplet degeneracy is lifted. This is due to the fact that, as the interaction is turned on, states with highest possible symmetry in the spin part of the wave function are favored because that way one gains the exchange energy. As a result, the triplet state comes down in energy with respect to others and therefore the period and the amplitude of the ground state

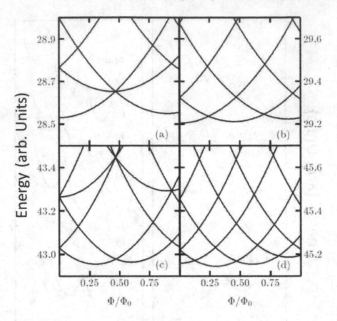

FIGURE 4.11
Few low-lying energy states for a ring containing two (a) non-interacting
and (b) interacting electrons. Three electron results are shown in (c) and
(d) for non-interacting and interacting systems, respectively.

oscillations is halved. Similar results are also seen for three electrons
$(S_z = \frac{1}{2})$ (Fig. 4.11 (c) non-interacting and (d) interacting electron systems), where the ground state oscillates with a period $\Phi_0/3$. Similar
behavior for ts also found for the QRs containing up to four electrons.
In the case of four electrons, in the absence of the Zeeman energy, the
spin configuration $S_z = 0$ has the lowest energy and a $\Phi_0/2$ periodicity.
However, if the Zeeman energy is included, the $S_z = 1$ configuration
becomes lower in energy and the $\Phi_0/4$ periodicity is observed.

The final message that emerges from these studies is the following:
the Coulomb interaction (in fact, any type of repulsive interactions) favors the spin-triplet ground states. In the absence of any interaction, the
ground states are spin singlets and as a function of Φ/Φ_0 are parabolas
with minimum at about the integer values (exactly at integer values for
an ideal ring). When a repulsive interaction is turned on, singlet states
rise in energy more than the triplet state and for strong enough repulsion,

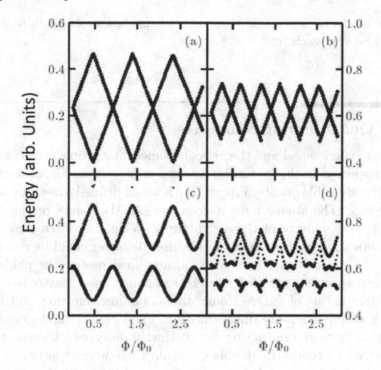

FIGURE 4.12

Absorption spectra of a QR containing two non-interacting [(a), (c)] or interacting [(b) and (d)] electrons. In (c) and (d), additionally, the system also has an impurity. As usual, the size of the filled circles is proportional to the calculated absorption intensities.

a decrease in the oscillation period is observed. Keyser et al.[22] reported transport spectroscopy on a small ring containing less than ten electrons. The deduced energy spectrum was found to be strongly influenced by the electron-electron interaction. They also observed a reduction of the AB period that is in line with our theoretical predictions.

Can one observe fractional oscillations of the ground state energy in optical spectroscopy? Our theoretical results for the optical absorption in a QR containing two electrons and also an impurity are shown in Fig. 4.12. The absorption spectra clearly reflect the behavior of the energy levels and the impurity does not destroy the fractional periodicity

[22]Fractional Aharonoc-Bohm oscillations in a Kondo correlated few-electron quantum ring, by U.F. Keyser, et al., *Adv. in Solid State Phys.* **43**, 113 (2003).

of the electronic states. Experimental confirmation of the above results would be very exciting.

4.5 Quantum ring complexes

Several experimental and theoretical studies on multiple quantum rings have appeared in the literature. Kuroda et al. created GaAs single and concentric double quantum rings[23] with a well defined inner ring and an outer ring. The atomic force microscope (AFM) images revealed clear rotational symmetry of these structures. What was interesting about the fabrication process was that the outer diameter could be controlled, while the inner diameter remained almost unchanged. The photoluminescence spectra[24] revealed discrete resonance lines that reflected the quantized nature of the electronic states of a quantum ring. In the concentric double quantum rings, emission originating from the outer ring and that from the inner ring are distinctly observed. Transport measurements on concentric double quantum rings were reported by Haug et al.[25]. The data revealed two different kinds of oscillations. One was identified as the expected Aharonov-Bohm effect from the outer ring. The other oscillation was attributed to the influence of charge redistributions in the inner ring. Photoluminescence of GaAs/AlGaAs triple concentric quantum rings[26] has also been reported. Few-electron eigenstates of concentric double quantum rings have been studied theoretically[27] by the exact-diagonalization scheme. At high magnetic fields, the energy spectrum is decoupled into spectra of inner and outer rings. Optical

[23]Self-Assembly of concentric quantum double rings, by T. Mano et al., *Nano Lett.* **5**, 425 (2005).

[24]Optical transitions in quantum ring complexes, by T. Kuroda, et al., *Phys. Rev. B* **72**, 205301 (2005).

[25]Coupling in concentric double quantum rings, by A. Mühle, W. Wegscheider and R.J. Haug, *Appl. Phys. Lett.* **91**, 133116 (2007); Coulomb-coupled concentric quantum rings, by A. Mühle, W. Wegscheider and R.J. Haug, *Physica E* **40**, 1246 (2008).

[26]Micro-photoluminescence of GaAs/AlGaAs triple concentric quantum rings, by M. Abbarchi et al., *Nanoscale Research Letters* **6**, 569 (2011).

[27]Few-electron eigenstates of concentric double quantum rings, by B. Szafran and F.M. Peeters, *Phys. Rev. B* **72**, 155316 (2005).

response of few-electron concentric double quantum rings have also been reported[28].

4.6 Rashba spin-orbit coupling revisited

The major incentive for including the spin-orbit interaction (SOI) in quantum rings is to have spin current in addition to the charge flow in the persistent current. One such system that has been proposed, is a spin-interference device (Fig. 4.13). In the Aharonov-Bohm (AB) ring, i.e., the magnetic flux passes through the ring, but the electrons move in a field-free region, introduction of a uniform SOI would lead to the phase difference between the spin wave functions traversing in the two arms of the ring. This is because the electron spins precess (due to the Rashba SOI) in opposite directions between the clockwise and counterclockwise motions in the ring. The relative difference in spin precession angle at the interference point causes the phase difference. The Rashba SOI can be controlled by a gate electrode which covers the entire AB ring (see Fig. 4.13) which will also control the interference. A large conductance modulation will signify the expected spin interference. The current modulation is an essential aspect of spin based electronics[29].

There have been several theoretical studies involving the Rashba SOI effects on quantum rings. The SOI was found to strongly influence the persistent current and is dependent on the strength of the Rashba potential[30]. Spin-orbit coupling was reported to result in finite persistent spin current for even number of electrons in the ring[31]. Detailed theoretical studies of persistent spin current and charge current in an ideal one-dimensional ring and also a two-dimensional hard-wall ring in a magnetic

---

[28]Optical response of two-dimensional concentric double quantum rings: A local-spin-density-functional theory study, by F. Malet et al., *Phys. Rev. B* **74**, 193309 (2006).

[29]Electrical manipulation of spins in the Rashba two dimensional electron gas systems, by J. Nitta et al., *J. Appl. Phys.* **105**, 122402 (2009); Spin-transistor electronics: An overview and outlook, by S. Sugahara and J. Nitta, *Proc. IEEE* **98**, 2124 (2010).

[30]Effect of the spin-orbit interaction on persistent currents in quantum rings, by A.V. Chaplik and L.I. Magarill, *Supperlattices and Microstructures*, **18**, 321 (1995).

[31]Persistent current in ballistic mesoscopic rings with Rashba spin-orbit coupling, by J. Splettstoesser, M. Governale, and U. Zülicke, *Phys. Rev. B* **68**, 165341 (2003).

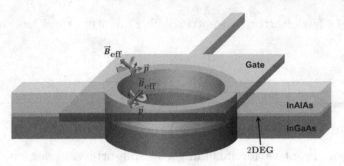

FIGURE 4.13
Schematic of spin interference device. The system has a strong spin-orbit interaction. The AB-ring area is covered with the gate electrode which controls the spin-orbit interaction.

field, in the presence of spin-orbit interactions were reported[32]. Quantum rings subjected to a Rashba SOI and an external magnetic field reveals unusual magnetic properties[33]. These authors also proposed a direct route to measure the SO coupling strength. Rashba SOI was also shown to create a remarkable change in the Raman spectrum of a few-electron quantum ring[34] due to the AB oscillations. Theoretical studies of spin-dependent magnetotransport subject to SO couplings were also reported[35].

---

[32]Spin states and persistent currents in mesoscopic rings: Spin-orbit interactions, by J.S. Sheng and Kai Chang, *Phys. Rev. B* **74**, 235315 (2006).

[33]Some unique properties of nanoscale quantum rings subjected to a Rashba spin-orbit interaction, by Hong-Yi Chen, P. Pietiläinen, and T. Chakraborty, *Phys. Rev. B* **78**, 073407 (2008).

[34]Effect of the spin-orbit coupling on the Raman spectra of a GaAs quantum ring with few electrons, by A. Manaselyan, A. Ghazaryan, and T. Chakraborty, *Solid State Commun.* **181**, 34 (2014).

[35]Spin-dependent magnetotransport through a mesoscopic ring in the presence of spin-orbit interaction, by X.F. Wang and P. Vasilopoulos, *Phys. Rev. B* **72**, 165336 (2005); Spin interference effects in ring conductors subject to Rashba coupling, by D. Frustaglia and K. Richter, *Phys. Rev. B* **69**, 235310 (2004).

4.7 Quantum ring and topological charge

We present below some interesting topological features of the spin fields in quantum rings (see Footnote 32) subjected to Rashba spin-orbit coupling (SOC) described above and additionally the Dresselhaus SOC[36] also being present. The corresponding Hamiltonians are: for the Rashba SOC, $H_R = \alpha(\sigma_x p_y - \sigma_y p_x)$, where the σs are, as usual, the Pauli spin matrices and $p_{x,y}$ are the components of the planar momentum. For the linear Dresselhaus SOC, $H_D = \beta(\sigma_y p_y - \sigma_x p_x)$. Here α, β are the coupling strengths of the SOCs. The in-plane spin field is defined as a vector field $(\sigma_x(\mathbf{r}), \sigma_y(\mathbf{r}))$ of which the two components are defined as $\sigma_s(\mathbf{r}) = \Psi^\dagger(\mathbf{r})\sigma_s\Psi(\mathbf{r})$ with the single-electron wave function $\Psi(\mathbf{r})$.

The topological index (charge) of the spin field[37] can be defined by the winding number of the spin field on an enclosed curve [in QDs (Sec. 3.6), the curve is a closed contour around the origin]

$$q(r) = \frac{1}{2\pi}\oint_r \frac{\sigma_x(\mathbf{r})\,d\sigma_y(\mathbf{r}) - \sigma_y(\mathbf{r})\,d\sigma_x(\mathbf{r})}{[\sigma_x(\mathbf{r})]^2 + [\sigma_y(\mathbf{r})]^2},$$

where the contour is selected as a circle with radius r. The topological charge of the spin field in a quantum ring is naturally defined by $q(r_0)$ with the radius of the ring r_0, since the density of the electron is located in the ring and the topological properties associated with the electron density are more meaningful and interesting.

The topological charge is $q = 1$ if only the Rashba SOC is present, while $q = -1$ if only the Dresselhaus SOC is present. When both SOCs are present at the same time the topological charge varies periodically with the increase of the magnetic field. The periodical topological transition of the spin field is primarily associated with the Aharonov-Bohm (AB) effect where the period is defined by increasing a magnetic flux quantum threading the area enclosed by the ring. However, the period of the topological transition is not exactly fixed by the AB period but is slightly corrected by the radial distribution of the electron density. The electron density is largely distributed along the ring, except being slightly shifted from the ring. By introducing an effective radius r_0' of

---

[36]Spin-Orbit Coupling Effects in Zinc Blende Structures, by G. Dresselhaus, *Phys. Rev.* **100**, 580 (1955).

[37]Magnetic field controlled topological transitions of the spin field in quantum rings with spin orbit couplings, by S. Peng, et al., *Physica E* **128**, 114545 (2021).

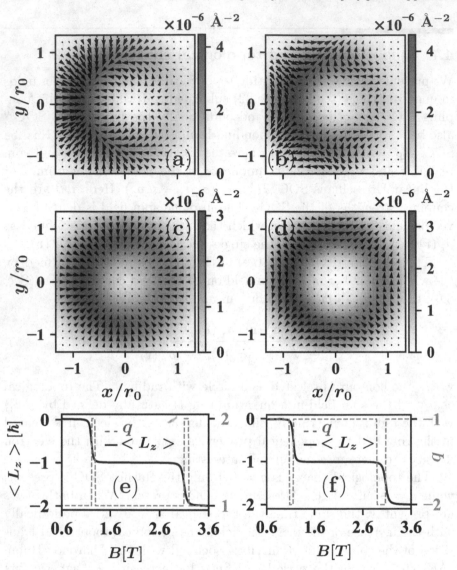

FIGURE 4.14

The in-plane spin textures of the InAs ring, with $r_0 = 30$ nm, $\hbar\omega = 10$ meV, with an impurity $\gamma = 0.02$ and $\delta = 0.3$. Spin textures of the ground state with only the Rashba SOC ($\hbar\alpha = 10.0$ nm·meV) for (a) $B = 1.25$ Tesla and for (c) $B = 1.05$ Tesla. The in-plane spin textures of the ground state with only the Dresselhaus SOC ($\hbar\beta = 10$ nm· meV for (b) $B = 3.0$ Tesla and for (d) $B = 3.5$ Tesla. The grey scale represents the density of the electron. The relation between q (the topological charge) and $\langle L_z \rangle$ for different magnetic fields, (e) for the Rashba SOC, and (f) for the Dresselhaus SOC.

the electron, $r_0' = \int \Psi^\dagger(\mathbf{r}) r \Psi(\mathbf{r}) \, d\mathbf{r}$ which varies with different magnetic fields and the SOCs, and is biased from the radius of the ring r_0, the fluctuation of the period of the topological transition can be understood to a certain extent by the effective magnetic flux threading the circle of the electron density with radius r_0'.

One interesting aspect of this study with a non-symmetric ring (i.e., a ring with a non-magnetic impurity in it) indicates that the impurity in the ring can easily change the topological charge of the single-electron spin field up to ± 2. The impurity is introduced as an external potential into the ring system with the Hamiltonian:[38]

$$H_{\text{imp}} = \tfrac{1}{2} m^* \omega^2 r_0^2 \gamma \, e^{-\frac{(x-r_0)^2 + y^2}{\delta^2 r_0^2}},$$

where γ and δ are tunable dimensionless parameters to characterize the size of the impurity effectively, and the impurity is supposed to be located at $(r_0, 0)$.

As the impurity is non-magnetic it does not directly couple to the electron spin. However, it changes the electron density directly and leads to the associated spin textures redistributed, especially when the angular momentum of the electron $\langle L_z \rangle$ is sharply varied. This is similar to the case of a quantum dot helium where the Coulomb interaction plays the role to change the density profile and the spin textures[39]. To see this point more clearly, we explore the curve of $\langle L_z \rangle$ versus the magnetic field in Figs. 4.14 (e) and 4.14 (f). Clearly, $\langle L_z \rangle$ varies stepwise but smoothly, since L_z does not commute with the Hamiltonian in the presence of the SOC and the impurity. On the plateau of $\langle L_z \rangle$, the topological charge is ± 1 as shown in Figs. 4.14 (c) and 4.14 (d). In the area between two neighboring plateaus, $q(r_0)$ can, in fact, be tuned to ± 2 as shown in Figs. 4.14 (a) and 4.14 (b), and the density profile is significantly biased from the rotational symmetric circle to make the spin textures change topologically. The mechanism is similar to the case in quantum dot helium, albeit the topological charge there is consistently ± 1. The tunable topological features of the spin field make the quantum ring suitable

---

[38]Persistent currents in a quantum ring: Effects of impurities and interactions, by T. Chakraborty and P. Pietiläinen, *Phys. Rev. B* **52**, 1932 (1995).

[39]Tuning the topological features of quantum-dot hydrogen and helium by a magnetic field, by W. Luo and T. Chakraborty, *Phys. Rev. B* **100**, 085309 (2019).

for topological spintronics[40] and the information stored as topological charge will perhaps be useful in quantum computation[41].

4.8   Rings in novel systems

In Sec. 2.9, we introduced the 2DEG at the oxide heterojunctions and some of its unique properties. ZnO is also found to be a perfect material to create quantum nanostructures. Preparation of nanorings, nanobelts, nanowires, etc. have been already reported[42]. We have already stated in Sect. 2.9 that the electrons are strongly correlated in these systems due to the increased electron effective mass and reduced dielectric constant of ZnO. Because of the enormous potential of this newly developed source of 2DEG, it is important that the electronic properties of quantum confined systems at the oxide interfaces are properly understood.

The low-lying energy levels of the ZnO quantum ring (QR), modelled as hard-core confinement potential, and containing one and two electrons (interacting) are described below. The results are also compared with those for the GaAs quantum rings. In a QR containing a single electron in both ZnO and GaAs systems the ground state energy changes periodically with increasing magnetic field [Fig. 4.15], as expected. In the case of ZnO QR the energy eigenvalues are lower than those in GaAs, due to the larger values of the electron effective mass. Additionally, the states with different spins are highly split due to the larger values of the g-factor in ZnO.

For the systems containing two interacting electrons, there are major differences in the energy levels between the two systems [Fig. 4.16]. For instance, in the GaAs QR, we notice the usual and well defined periodic oscillations due to the level crossings between the singlet and triplet ground states. For each crossing of the two-electron ground state, the total angular momentum L changes by unity. On the other hand, for the ZnO QR containing two interacting electrons, due to the combined effect of the strong Zeeman splitting and the strong Coulomb interaction,

---

[40]Spintronics: Fundamentals and applications, by I. Zutic, J. Fabian, and S. Das Sarma, *Rev. Mod. Phys.* **76**, 323 (2004); Topological antiferromagnetic spintronics, by L. Smejkal, et al., *Nat. Phys.* **14**, 242 (2018).

[41]Arbitrary qubit transformations on tuneable Rashba rings, by A. Kregar, J.H. Jefferson, and A. Ramsak, *Phys. Rev. B* **93**, 075432 (2016).

[42]Nanostructures of zinc oxide, by Z.L. Wang, *Materials Today* **7**, 26 (2004).

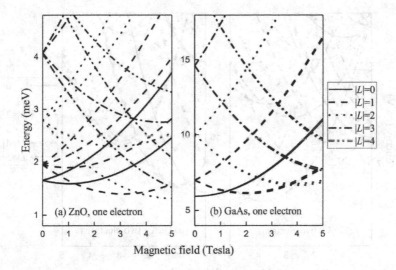

FIGURE 4.15
(a) Low-lying energy levels versus the magnetic field for the ZnO quantum ring containing a single electron. (b) The same for a GaAs QR.

the singlet-triplet crossings are no longer present in the ground state. Interestingly, those periodic crossings can be noticed only in the excited state. For a small value of the magnetic field, the ground state is a singlet with $L = 0$ and the total electron spin $S = 0$. With an increase of the magnetic field the ground state changes to a triplet state with $L = -1$ and $S = -1$. With a further increase of the magnetic field, all the observed crossings of the ground state correspond to triplet-triplet transitions between the states with odd number of total angular momentum ($|L| = 1, 3, 5, ...$). These interesting and unusual results will manifest itself as unusual behaviors in optical transitions in the ZnO QR.

The optical transitions corresponding to the energy spectra discussed above reveal some very unusual behavior for the QR in ZnO. The transitions are shown for ZnO in Fig. 4.17, and in Fig. 4.18 for the conventional GaAs quantum ring. As before, the size of the points are proportional to the calculated intensity of the optical transitions. For the rings containing only a single electron in both systems (ZnO and GaAs), we notice the expected periodic oscillations. We also notice that while the strong

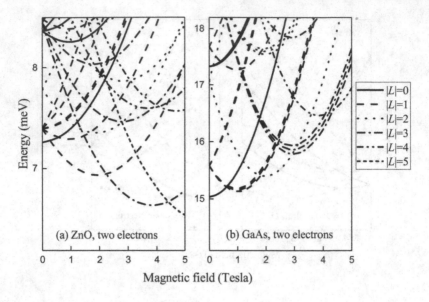

FIGURE 4.16
(a) Low-lying energy levels versus the magnetic field for the ZnO quantum ring containing two interacting electrons. (b) The same for a GaAs QR.

Zeeman energy changes the one-electron spectra of the QR in ZnO, it does not alter the periodicity of the oscillations in the optical spectra.

The situation is however different for quantum rings with two interacting electrons. In the case of a GaAs ring, the oscillations appear with a period that is half the flux quantum, which is a well known result (see Sec. 4.4). In stark contrast, for the rings in ZnO we notice a 'aperiodic' behavior of the oscillations: The first oscillation which corresponds to the singlet-triplet transitions from the $L = 0$ state to the $L = -1$ state has a smaller period compared to the other oscillations that correspond to transitions between the triplet states with odd angular momentum. The period of those triplet-triplet transitions is almost equal to that of the single-electron case (and not period halving, as expected). This unusual behavior can be explained by the combined effect of the strong Zeeman interaction and the strong electron-electron interaction in ZnO.

The low-lying energy levels for the rings in ZnO and those in GaAs, containing three interacting electrons are shown in Fig. 4.19 (a) and (b), as a function of the magnetic field. For the GaAs ring we notice

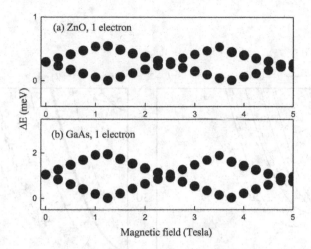

FIGURE 4.17

(a) Optical transition energies versus the magnetic field, for a ZnO quantum ring containing one electron. (b) the same for a GaAs ring. The size of the dots is proportional to the intensity of the calculated optical transitions.

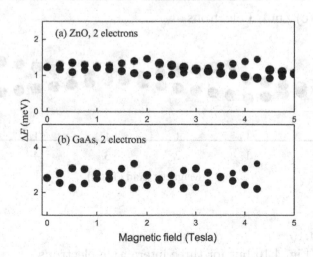

FIGURE 4.18

Same as in Fig. 4.17, but for rings containing two interacting electrons.

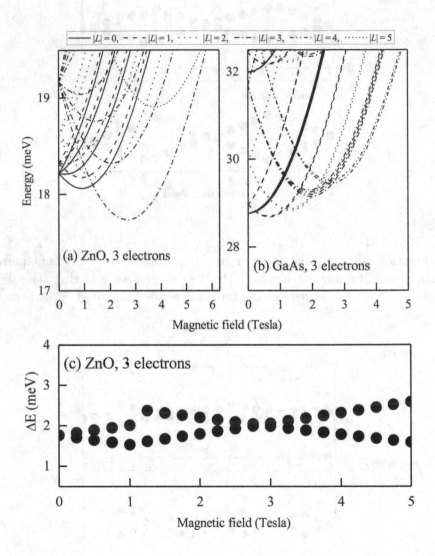

FIGURE 4.19
Same as in Fig. 4.16 but for three interacting electrons.

the periodic ground-state transitions and during each transition, the ground-state angular momentum changes by unity. Interestingly, for the ZnO system, the situation is entirely different. for a three-electron QR in ZnO, in that magnetic field range, only two transitions are present.

At low magnetic fields the ground state has the $L = 0$ angular momentum. As the magnetic field increases to $B = 1.3$ Tesla the ground state changes to $L = -3$. The next ground state transition happens at $B = 6$ Tesla and the angular momentum changes to $L = -5$. It is therefore clear that within the range of magnetic field considered here, the periodicity of the energy spectra disappears. The disappearence of the periodic behavior is clearly visible in the optical transition energies (Fig. 4.19 (c)). The persistent current, that is always associated with a quantum ring is found to be highly dependent on the number of electrons in a ZnO quantum ring and could be controlled externally. This property could prove to be beneficial in future practical applications of the novel system of ZnO quantum rings.

4.9   Isotropic or anisotropic?

Anisotropy in quantum rings, be it the effective-mass anisotropy, the shape anisotropy, or anisotropy associated with defects, strongly influence the electronic, magnetic, or optical properties of a quantum ring[43]. Anisotropy can, in general, drastically alter the periodic oscillations of the energy spectra and thereby influence the persistent current. That is certainly not a great news if you are to study the shape dependence of the electronic properties of quantum rings of any shape. Interestingly, theoretical studies indicate that under the influence of intense terahertz laser fields, the electronic and optical properties of quantum rings undergo dramatic changes. Theoretical studies of the energy spectra and optical transitions of circular and elliptical QRs clearly confirmed that, in the case of isotropic QRs, the laser field can create very unusual behavior of the energy spectra that are characteristics of anisotropic rings.

---

[43]Electron states in quantum rings with structural distortions under axial or in-plane magnetic fields, by J Planelles, F Rajadell and J I Climente, *Nanotechnology* **18**, 375402 (2007); Unusual quantum confined Stark effect and Aharonov-Bohm oscillations in semiconductor quantum rings with anisotropic effective masses, by G.O. de Sousa, et al., *Phys. Rev. B* **95**, 205414 (2017).

Similarly, in the case of anisotropic rings, the laser field can *restore* the isotropic behavior. Therefore, with the help of the laser field, it is theoretically possible to create a continuous evolution of the energy spectra and optical properties of structurally anisotropic quantum rings to those of isotropic rings in a controlled manner in a single ring[44]. Experimental confirmation of this interesting finding would provide a very important step forward in our exploration of the physical properties of quantum rings.

4.10 Device applications of the quantum rings

What good are the quantum rings for quantum devices? Quantum rings have found applications mostly in optoelectronic devices. In(Ga)As quantum rings have been used as terahertz detectors[45]. In(Ga)As quantum ring photodetectors have been fabricated to operate in the spectral range of 0.4-6.0 μm[46]. In the presence of a lateral electric field exceeding a certain threshold, it is shown[47] that one can switch the ground state of an electron-hole pair in a quantum ring from being optically active (bright) to optically inactive (dark). A possible application could be exciton storage and readout. Other applications include, quantum ring based solar cells[48], and single-photon emitter with antibunching features (see Sec. 3.7.2)[49]. Quantum rings with spin-orbit couplings can be used as a spin analyzer by studying the charge transmission rate of the

---

[44]Controllable contonuous evolution of electronic states in a single quantum ring, by T. Chakraborty, A. Manaselyan, M. Barseghyan, and D. Laroze, *Phys. Rev. B* **97**, 041304 (R) (2018).

[45]In(Ga)As quantum rings for terahertz detectors, by J.-H. Dai et al., *Jap. J. Appl. Phys.* **47**, 2924 (2008); High-performance quantum ring detector for the 1-3 terahertz range, by S. Bhowmick, et al., *Appl. Phys. Lett.* **96**, 231103 (2010).

[46]Multicolor photodetector based on GaAs quantum rings grown by droplet epitaxy, by J. Wu, et al., *Appl. Phys. Lett.* **94**, 171102 (2009).

[47]Exciton storage in a nanoscale Aharonov-Bohm ring with electric field tuning, by Andrea M. Fischer, et al., *Phys. Rev. Lett.* **102**, 096405 (2009); Aharonov-Bohm interference in neutral excitons: Effects of built-in electric fields, by M.D. Teodoro, et al., *Phys. Rev. Lett.* **104**, 086401 (2010).

[48]Strain-free ring-shaped nanostructures by droplet epitaxy for photovoltaic application, by J. Wu, et al., *Appl. Phys. Lett.* **101**, 043904 (2012).

[49]Photon antibunching in double quantum ring structures, by M. Abbarchi, et al., *Phys. Rev. B* **79**, 085308 (2009).

incident spin polarization. A detailed proposal to design such an isotropic all electrical spin analyzer with a quantum ring containing spin-orbit couplings has been recently reported[50].

---

[50]Isotropic all-electric spin analyzer based on a quantum ring with spin-orbit couplings, by S. Peng, et al., *Appl. Phys. Lett.* **118**, 082402 (2021).

5

Graphene: Carbon and its nets

Graphene is a single layer of graphite. Therefore, we begin this chapter with an introduction of the properties of the fascinating mineral – the graphite. Graphite (the Plumbago, or Black Lead as it was known in the old days) is a crystalline form of carbon. It is found naturally in its mineral form, but can also be produced by synthetic processes. This black lustrous mineral has been used by humans for centuries. In the sixteenth century, a huge deposit of graphite was discovered in England and soon its industrial (and military) potentials were realized, especially, in making moulds for casting perfect smoother, rounder cannon balls, which travelled farther thus allowing the British Navy to rule the waves. Use of graphite as pencils happened around 1565. Today, graphite is used in a vast array of important applications. One of the most useful properties of graphite is that it maintains its strength and stability at extremely high temperatures. Hence, it is suitable in refractory (heat-resistant) applications, that involves use of extremely high heat. It is used in rocket nozzles, brake linings, and for many other purposes. It is also used as a moderator in nuclear reactors. Graphite is an electrical conductor. In fact, it is the only non-metal that conducts electricity. It is used as dry lubricant in machine parts. These are only a few uses mentioned here from the long list of useful industrial applications of this extremely versatile mineral[1]. In 1789, a German chemist and mineralogist Abraham Gottlob Werner named the mineral 'Graphite' that was derived from the Greek verb 'Graphein' meaning 'to write'. In the early days of the discovery of graphite, it was purportedly used for marking sheep by the farmers.

---

[1] *Carbon Materials: Science and Applications*, by D.D.L. Chung (World Scientific, Singapore 2019).

DOI: 10.1201/9781003090908-5

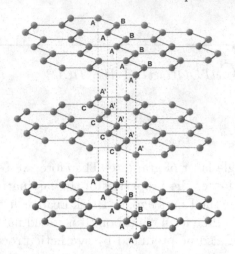

FIGURE 5.1
The structure of graphite according to Bernal.

5.1   A brief history of graphene

The history of graphene research, with many ground-breaking contributions prior to 2004, has been largely forgotten in the rush to move forward, as has been pointed out by several authors[2]. Our present-day understanding and an unusual fascination with graphene (more of that later) did not happen in a vacuum. Most of the foundations related to our current knowledge of graphene had been painstakingly laid down by a few scientists (both theoretical and experimental) who deserve to be properly recognized. The contributions by those pioneers in graphite/graphene physics were truly phenomenal. In fact, from that perspective, the modern-day discoveries of physical phenomena related to graphene, though immensely popular, do not really constitute a

---

[2]From conception to realization: An historical account of graphene and some perspectives for its future, by D.R. Dreyer, et al., *Angew. Chem. Int. Ed.* **49**, 9336 (2010); The electrochemistry of CVD graphene: progress and prospects by D.A.C. Brownson and C.E. Banks, *Phys. Chem. Chem. Phys.* **14**, 8264 (2012); Nobel document triggers debate, by E.S. Reich, *Nature* **468**, 486 (2010).

'paradigm shift'[3]. We first describe those earlier studies before we embark upon our trip to the modern-day work on graphene.

5.1.1 Major breakthroughs in graphene research

One major step in our understanding of the properties of graphene began with the study of the structure of graphite reported in the classical work by J.D. Bernal in 1924. He determined the structure of graphite by means of X-ray diffraction analysis from a rotating crystal. According to Bernal, graphite consists of parallel layers, each of which is hexagonal nets (say in the xy plane)[4] of carbon atoms, in which the atoms are separated by 1.42 Å. The layers are then placed on top of each other in a sequence 1-2-1-2 ..., which means that alternate layers have same projections on the xy plane. In other words, the Bernal structure can be visualized as hexagonal nets of carbon atoms stacked in layers so that half the atoms in one layer are directly above those in the layer below; the other half being above the holes in the first layer (Fig. 5.1). The interlayer separation is 3.35 Å and the unit cell contains four carbon atoms. This is the most common (and stable) stacking of layers in graphite.

Box 5.1 Crystal structure and electronic properties of graphene

There are some excellent books[5,6,7], that have been published where the details about the crystal structure, the band structure and many other physical properties of graphene are available. Graphene is a honeycomb lattice of carbon atoms. It is a bipartite lattice with two sublattices A and B that are triangular lattices. Considering only the xy-plane, the unit vectors in real space $\mathbf{a}_1$ and $\mathbf{a}_2$, and in the reciprocal lattice[8] vectors $\mathbf{b}_1$ and $\mathbf{b}_2$ are shown in Fig. 5.2 (a), (b). The first Brillouin zone (primitive cell in reciprocal space) is a hexagon, as shown where the corners are called the K points. The six corners form two inequivalent groups of K points, traditionally labeled as K and K'. As described below, these points

---

[3] *The Structure of Scientific Revolutions*, by Thomas S. Kuhn, The University of Chicago, 1970.

[4] *The Structure of Graphite*, by J.D. Bernal, Proc. Roy. Soc. A **106**, 749 (1924).

are also called the 'Dirac points'.

The next major advancement took place in 1947 when the band structure of graphite was reported by Wallace[9] and independently by Coulson[10]. These authors made a simplifying approximation and considered only a single layer of graphene to represent graphite. Electrons are then confined within a single layer of graphene. The most appropriate method to study the electron dispersion is the tight-binding approximation. The energy dispersion relation in the tight-binding approximation is shown in Fig. 5.2 (c) (in the nearest neighbor approximation)[11] as a function of the wave vector $\mathbf{k}$. The upper half of the curves is called the π^* or the anti-bonding band while the lower one is π or the bonding band. The two bands meet at the K points where the dispersion vanishes. This is also the Fermi energy level.

As there are two electrons per unit cell they fully occupy the lower π band. The first Brillouin zone (the primitive cell in reciprocal, or momentum space) is hexagonal and at two of its inequivalent corners (the K and K' points), the conduction and valence bands overlap. Graphene is often described as a two valley (K and K') zero-bandgap semiconductor. Because the filled and empty bands overlap, even an extremely small amount of thermal energy would be sufficient to excite copious electrons from the filled (valence) band to the empty (conduction) band. Consequently, at any temperature above absolute zero there are free electrons in the conduction band, thereby rendering graphite an extremely good conductor within the plane of graphene. The electrical conductivity of graphite is highly anisotropic: the conductivity in the graphene plane is

---

[5]*Physics of Graphene*, edited by H. Aoki and M.S. Dresselhaus (Springer, New York 2014).

[6]*Graphene Nanoelectronics*, edited by H. Raza (Springer, New York 2012).

[7]*Graphene: Carbon in Two Dimensions*, by M.I. Katsnelson (Cambridge University Press, New York 2012).

[8]Fourier transform of the real space lattice, where the lattice vector has units of wave vector $\mathbf{k}$.

[9]The Band Theory of Graphite, by P.R. Wallace, *Phys. Rev.* **71**, 622 (1947).

[10]Energy Bands in Graphite, by C.A. Coulson, *Nature* **159**, 265 (1947); Studies in Graphite and Related Compounds I: Electronic Band Structure in Graphite, by C.A. Coulson and R. Taylor, *Proc. Phys. Soc. A* **65**, 815 (1952).

[11]For a complete and detailed description of the properties of graphene, see, e.g., Properties of graphene: a theoretical perspective, by D.S.L. Abergel, et al., *Adv. Phys.* **59**, 261 (2010).

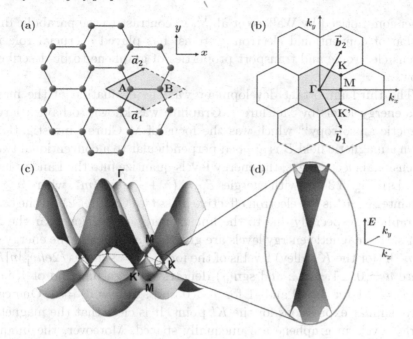

FIGURE 5.2
(a) Graphene lattice in real space. The corresponding reciprocal lattice. The unit cell of graphene contains two atoms A and B. The first Brillouin zone is shown as shaded hexagon. The basis vectors of the direct lattice and the reciprocal lattice are a_i and b_i respectively. The high-symmetry points Γ, M and K in the Brillouin zone are also indicated. (c) Energy dispersion in graphene. Since we ignore the coupling between the graphene sheets, the energy bands depend only on k_x and k_y. The lower band (the π band) is completely filled and meets the totally empty upper band (the π^* band) at the K and K' points. Near these points both bands have linear dispersion as described in the text.

'at least ten thousand times larger than that along the normal to the plane'[12].

Interestingly, near the K points, the energy dispersion is well approximated by $\mathcal{E}_\pm(k) = \pm \hbar v_F |\mathbf{k}|$, where $\mathbf{k}$ is measured with respect to K (or K') point and $v_F = 10^6$ m/s is the Fermi velocity. The *linear*

---

[12]Large anisotropy of the electrical conductivity of graphite, by K.S. Krishnan and N. Ganguli, *Nature* **144**, 667 (1939).

dispersion, noticed by Wallace et al.,[13], in contrast to the parabolic dispersion of conventional electron systems, has played a crucial role in various electronic and transport properties of graphene to be described below.

The third important development was the evaluation of the magnetic energy levels by McClure[14]. Graphite was known to have a large magnetic anisotropy[15] which was the focus of McClure's investigation. When a magnetic field B is applied perpendicular to a conventional two-dimensional electron gas, the energy levels quantize into the Landau levels (LLs) [Fig. 5.3 (a)] with energies $\mathcal{E}_n = \left(n + \frac{1}{2}\right) \hbar e B / m^*$, where $n \geq 0$ is an integer, m^* is the electron effective mass (see Chapter 2). In the case of graphene, especially due to the linear energy dispersion near the K points, the magnetic energy levels are greatly modified and the energy of the n-th (for the K valley) level is of the form, $\mathcal{E}_n = \mathrm{sgn}(n)\sqrt{2e\hbar v_\mathrm{F}^2 |n| B}$, where $n = 0, \pm 1, \pm 2, ...$ and $\mathrm{sgn}(j)$ denotes the sign function of j, i.e., $\mathrm{sgn}(j) = +1$ for $j > 0$ and -1 for $j < 0$ and $\mathrm{sgn}(j = 0) = 0$. One can derive similar expressions for the K' point. It is clear that the magnetic energy levels in graphene are unequally spaced. Moreover, the unique feature here is that the level at $n = 0$ is *independent of the magnetic field strength* [see Fig. 5.3 (b)]. The $n = 0$ LL is shared between the valence and conduction band, and is half filled for pure graphene. McClure noted that the $n = 0$ levels are responsible for the most part of the large diamagnetism in graphite. In fact, the magnetization at $n = 0$ is $\mathcal{M} \propto -\sqrt{B}$ and the electrons in the $n = 0$ LL make the dominant contribution to graphene diamagnetism[16].

These are the theoretical foundations that made the rapid progress of the modern day research on graphene possible. Interestingly, there were also a few seminal experimental works reported on graphene, in particular, on the adhesive tape method of exfoliating graphene, that began as early as in the sixties.

---

[13]The electric and magnetic properties of graphite, by R.R. Haering and P.R. Wallace, *J. Phys. Chem. Solids* **3**, 253 (1957).

[14]Diamagnetism of graphite, by J.W. McClure, *Phys. Rev.* **104**, 666 (1956).

[15]Magnetic and other properties of the free electrons in graphite, by N. Ganguli and K.S. Krishnan, *Proc. Roy. Soc. Lond. A* **117**, 168 (1941); Temperature variation of the abnormal unidirectional diamagnetism of graphite crystals, by K. Krishnan and N. Ganguli, *Nature* **139**, 155 (1937).

[16]Field and temperature dependence of intrinsic diamagnetism in graphene: Theory and experiment, by Z. Li, et al., *Phys. Rev. B* **91**, 094429 (2015).

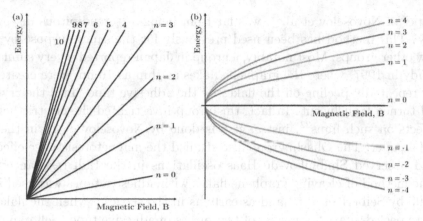

FIGURE 5.3
The magnetic energy levels for (a) electrons in conventional two-dimensional electron systems, and (b) in graphene.

5.1.2   Isolating graphene: Sellotape versus the Scotch tape

In contrast to the graphene folktales[17], graphene was actually not 'discovered' in 2004. There is a long and fruitful history of isolation of graphene that precedes that popularly claimed time period. Soon after the theoretical work of Wallace and Coulson were published, the first transmission electron micrograph of isolated graphene was reported by Ruess and Vogt[18] in 1948. They made graphene by thermal reduction of graphene oxide in a stream of hydrogen. Later in 1962, Boehm et al. isolated free graphene by reducing graphene oxide with hydrazine[19]. In fact, the credit for coining the name 'graphene' goes to Boehm et al.[20] in 1987. Most significantly, the 'scotch-tape' method of peeling off graphene flakes from bulk graphite blocks, popularized by the 2004

---

[17]For example, several news media such as, *Europhysics News* **40**, 17 (2009);
https://www.aps.org/publications/apsnews/200910/physicshistory.cfm
and countless other articles often begin by stating that graphene was discovered in 2004.

[18]Höchstlamellarer Kohlenstoff aus Graphitoxyhydroxyd..., by G. Ruess and F. Vogt, *Monatshefte für Chemie* **78**, 222 (1948).

[19]Dünnste Kohlenstoff-Folien, by H.P. Boehm, et al., Z. Naturforsch. B**17**, 150 (1962).

[20]Graphene – How a laboratory curiosity suddenly became extremely interesting, by Hans-Peter Boehm, *Angew. Chem. Int. Ed.* **49**, 9332 (2010).

paper of Novoselov et al.[21], was far from being a 'serendipitous discovery'. This method has been used previously for the same purpose by a few other groups. Most notably, a group in Japan reported a very similar study in 1997, where the graphene flakes as thin as 10 nm were created by repeatedly peeling off the flakes by the adhesive tape 'until the crystal turned translucent'. In fact, this group investigated the electric field effects on such films[22], just as it was done by Novoselov et al. in their 2004 paper. The Ohashi group also studied the magnetoresistance effect and observed Shubnikov-de Haas oscillations in the Hall resistance[23]. The method of cleaving graphene flakes with adhesive tapes was used in 1990 by Seibert et al.[24], and as early as in the sixties[25] when the flakes were peeled apart 'by means of two pieces of adhesive tape (Sellotape)'. It is not clear if in those adhesive tape experiments some flakes were actually a single-layer graphene. A monolayer graphene created by the 'scotch tape' method was indeed reported by Novoselov et al.[26] in 2005.

Irrespective of how it was made, isolating a single sheet of graphene was a major feat[27] that ushered in a new paradigm for novel two-dimensional electron systems having some spectacular properties which resulted in a deluge of experimental and theoretical activities on graphene worldwide.

---

[21]Electric field effect in atomically thin carbon films, by K.S. Novoselov et al., *Science* **306**, 666 (2004).

[22]Size effect in the in-plane electrical resistivity of very thin graphite crystals, by Y. Ohashi, et al., *TANSO*, 235 (1997).

[23]Magnetoresistance effect of thin films made of single graphite crystals, by Y. Ohashi, et al., *TANSO*, 410 (2000).

[24]Femtosecond carrier dynamics in graphite, by K. Seibert, et al., *Phys. Rev. B* **42**, 2842 (1990).

[25]The identification and some pseudo-chemical consequences of non-basal edge and screw dislocations in graphite, by C. Roscoe and J.M. Thomas, *Proc. R. Soc. Lond. A* **297**, 397 (1967).

[26]Two-dimensional atomic crystals, by K.S. Novoselov et al., *PNAS* **102**, 10451 (2005).

[27]*Advanced Materials Science and Engineering of Carbon*, by M. Inagaki, F. Kang, M. Toyoda, and H. Konno, (Elsevier 2014).

5.2 Electrons behaving differently

In the continuum limit and in the effective mass approximation, staying in the vicinity of the K point we could write the graphene Hamiltonian as $\mathcal{H}_K = \hbar v_F \begin{pmatrix} 0 & k_x - ik_y \\ k_x + ik_y & 0 \end{pmatrix}$, or equivalently, $\mathcal{H}_K^D = \hbar v_F \sigma \cdot \mathbf{k}$, where $\sigma = (\sigma_x, \sigma_y)$ is a vector of Pauli matrices: $\sigma_x = \begin{pmatrix} 0 & 1 \\ 1 & 0 \end{pmatrix}$ and $\sigma_y = \begin{pmatrix} 0 & -i \\ i & 0 \end{pmatrix}$. This Hamiltonian will have the linear energy dispersion $\mathcal{E}_\pm(k) = \pm v_F \hbar k$ which we have already discussed above. The Hamiltonian at the K' point is obtained by exchanging $\Pi_\pm = k_x \pm ik_y$ in the above equation. Interestingly, the Hamiltonian $\mathcal{H}_K^D$ is exactly that of *massless Dirac fermions*, just as for the neutrinos (the little neutrons in Italian, coined by Enrico Fermi)[28], in two spatial dimensions[29]. We should emphasize that in $\mathcal{H}^D$ we do not actually deal with the real spin but the 'pseudospin' that lets us differentiate between contributions from each of the two sublattices (see below for more on this).

The wave functions of the Dirac-like particles have a spinor structure (for K and K' points)

$$\psi_{s,k}^K = e^{i\mathbf{k}\cdot\mathbf{r}} \begin{pmatrix} s \\ e^{i\theta(k)} \end{pmatrix}; \quad \psi_{s,k}^{K'} = e^{i\mathbf{k}\cdot\mathbf{r}} \begin{pmatrix} e^{i\theta(k)} \\ s \end{pmatrix},$$

where $s = +1$ for the upper band and -1 for the lower band, $\tan\theta = k_y/k_x$, and the upper and lower terms correspond to the quantum mechanical amplitudes (or 'pseudospin') of finding the particle on one of

---

[28] A classical relativistic free particle with mass m and momentum $\mathbf{p}$ has energy $E = (p^2c^2 + m^2c^4)^{\frac{1}{2}}$. In the ultra-relativistic regime, i.e., $cp \gg mc^2$, the dispersion becomes linear. For a spin-$\frac{1}{2}$ particle that is confined to the plane $\mathbf{r} = (x, y)$, a quantum Hamiltonian whose square gives E^2 is $\mathcal{H}_D = c\,\sigma \cdot \mathbf{p} + mc^2\sigma_z$, where $\sigma_z = \begin{pmatrix} 1 & 0 \\ 0 & -1 \end{pmatrix}$. This is the Dirac Hamiltonian in 2D. The Hamiltonian acts on two-component spinors. In the case of graphene, we have $m = 0$. A massless particle is always in the ultra-relativistic regime. Photons have zero rest mass and move at the speed c, but they are spin-1 particles. They do not obey the Dirac equation, but the Maxwell equation. Neutrinos are long thought to be spin-$\frac{1}{2}$ particles with zero rest mass.

[29] Electronic states and transport in carbon nanotubes, by T. Ando, in *Nano-Physics & Bio-Electronics: A New Odyssey*, edited by T. Chakraborty, F. Peeters, and U. Sivan (Elsevier, Amsterdam 2002).

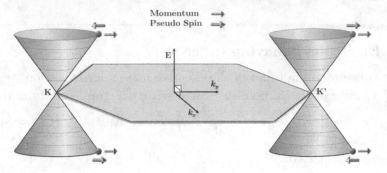

FIGURE 5.4
Helicity of electrons at the two Dirac cones.

the two sublattices, A and B. In graphene, the pseudospin direction is associated with the momentum of the particles. This means that the wave functions in the vicinity of the K and K' points are 'chiral'. The word 'chirality' means handedness (the particle is either left-handed or right-handed), and for particles of zero-mass it is equivalent to 'helicity'[30]. The helicity is the projection of spin σ onto the direction of motion: $h = \sigma \cdot \mathbf{q}/|\mathbf{q}|$. As the Dirac Hamiltonian at K can be written as $\mathcal{H}_K = \hbar v_\mathrm{F} \left(\sigma \cdot \mathbf{q} \right)/|\mathbf{q}|$, it is clear that $h_q \psi_\pm(\mathbf{q}) = \pm \psi_\pm(\mathbf{q})$. Simply stated, electrons (holes) in graphene have a definite pseudospin orientation, viz. parallel (antiparallel) to the direction of its momentum (Fig. 5.4). Particles have opposite chirality in the K and K' *valleys* or in the electron or hole bands. Helicity is a good quantum number in graphene, i.e., it is conserved. One consequence of this property is that, any intravelley backscattering, i.e., scattering of particles from states of opposite momentum and opposite pseudospin, is suppressed, even in the presence of a scatterer (long-range)[31].

It should however be stressed that the Dirac physics in graphene is merely an analogy between the two Hamiltonians shown above. There are no neutrinos whizzing around in graphene: the velocity of light is replaced here by the Fermi velocity which is a factor $\sim$300 smaller. Further, while the 'left-handed neutrino' in graphene is not equivalent to

---

[30]Dirac, Majorana, and Weyl fermions, by P.B. Pal, *Amer. J. Phys.* **79**, 485 (2011).

[31]Impurity scattering in carbon nanotubes – Absence of back scattering, by T. Ando and T. Nakanishi, *J. Phys. Soc. Jpn.* **67**, 1704 (1998); Berry's phase and absence of back scattering in carbon nanotubes, by T. Ando, T. Nakanishi, and R. Saito, *J. Phys. Soc. Jpn.* **67**, 2857 (1998).

the 'right-handed antineutrino' that lies at the point K' rather than at K, they are not really antiparticles of each other. Nevertheless, describing the low-energy properties of electrons in graphene in this framework reveals several intriguing properties to be described below.

The energy bands in the vicinity of the 'Dirac points' (K and K') are, $\mathcal{E} = \pm\hbar v_F k$, where $+(-)$ denotes the conduction (valence) band. Each energy level is four-fold degenerate due to two-fold spin and two-fold valley degeneracies. They look like two cones with their pointed tips meeting at $k = 0$ [Fig. 5.2 (d)]. The corresponding 'density of states' (the number of electron states available per unit volume per unit energy) varies linearly as a function of energy and vanishes at $\mathcal{E} = 0$. In contrast, in a conventional 2D electron system, the density of states is independent of energy. The (inverse) electronic compressibility which is related to the strength of electron-electron interaction[32] shows a very surprising behavior in monolayer graphene. In this system, the measured compressibility was found to be quantitatively described by the kinetic energy alone[33]. Theoretical studies by Abergel et al. indicated that there is total vanishing of the correlation contribution to compressibility in graphene due to the spinor structure of the single-particle wave functions, which is a direct manifestation of the chirality of electrons in monolayer graphene. In fact, it is the chirality that exhibits a very unique behavior in monolayer graphene: the electron correlations analytically vanish from the two-particle kinetic energy, something that has never been observed in conventional electron systems[34].

5.3 Quantum Hall effects in graphene

We have already discussed the nature of LLs in graphene. In the Dirac picture, the energies of the LLs depend only on the index n: $\mathcal{E}_n = \hbar\omega_B^D \mathrm{sgn}\sqrt{|n|}$, where $\omega_B^D = \sqrt{2}v_F/\ell_B$, and $\ell_B = \sqrt{\hbar/eB}$ is the magnetic length. The landau level index, n, can be positive or negative.

---

[32] *Many Particle Physics*, by G.D. Mahan, 3rd ed. (Kluwer/Plenum, New York 2000).

[33] Observation of electron-hole puddles in graphene using a scanning single-electron transistor, by J. Martin et al., *Nat. Phys.* **4**, 144 (2008).

[34] Electronic compressibility of graphene: The case of vanishing electron correlations and the role of chirality, by D.S.L. Abergel, P. Pietiläinen, and T. Chakraborty, *Phys. Rev. B* **80**, 081408 (R) (2009).

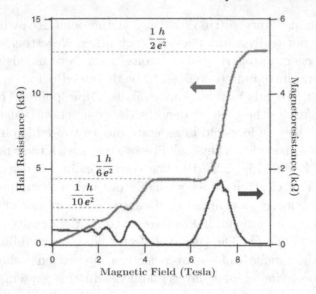

FIGURE 5.5

Quantum Hall effect in graphene (as observed by Kim et al.). The Hall resistance plateaus appear at the filling factors $\nu = 2, 6, 10$, just as expected for Dirac-like behavior of electrons in graphene.

The positive values correspond to electrons (conduction band), while the negative values correspond to holes (valence band). As already mentioned earlier, in contrast to the case of conventional 2DEG, the LLs in graphene are not equidistant and the largest energy separation is between the zero and the first LL. The energy difference between the lowest LL $(n = 0)$ and the next LLs $(n = \pm 1)$ is $\Delta E \approx 400\mathrm{K}\sqrt{B(\mathrm{Tesla})}$, which for $B = 20$ Tesla is 1800 K. Such a large electron energy gap allows one to observe the quantum Hall effect in graphene, even at room temperature[35].

Because of the unconventional nature of LLs, we expect a different type of quantum Hall effect, viz., the Hall conductivity should follow the relation, $\sigma_{xy} = \nu e^2/h = 2(2n + 1)e^2/h$, where n is an integer. In graphene, each LL has fourfold degeneracy due to valley and spin. The unique properties of the lowest LL $(n = 0)$ introduces a shift in quantization of the Hall conductance. This shift is in fact, related to the

---

[35]Room-Temperature Quantum Hall Effect in Graphene, by K.S. Novoselov et al., *Science* **315**, 1379 (2007).

electron-hole symmetry of the graphene layer. The zeroth LL $n = 0$ is half filled with electrons and half filled with holes of opposite chirality, when the system is charge neutral. This LL has zero energy regardless of the value of the magnetic field, and it has the properties of both electrons and holes. This quantum anomaly of the $n = 0$ LL makes this level effectively twofold degenerate for electrons and twofold degenerate for holes[36]. As a result, the quantization of the Hall conductance occurs at half-integer values as given above. This is precisely what was observed in the experiments[37]. Observation of the so-called 'half-integer QHE' is truly remarkable because, albeit a rediscovery of the quantum Hall effect in a new material, it actually confirms the Dirac-like behavior of electrons in graphene. Typical results obtained by the Kim group[38] are shown in Fig. 5.5.

Interestingly, for an ideal graphene system, the wave functions for an electron is characterized by a set of quantum numbers $\alpha = (j, n, k)$ where $j = K$ and K', the LL index $n = 0, \pm1, \pm2, ...$, and k is the electron wave vector, the direction of which is specified by the chosen gauge, and is expressed as[39]

$$\psi_{K,n} = C_n \begin{pmatrix} \mathrm{sgn}(n)\mathrm{i}^{|n|-1}\phi_{|n|-1} \\ \mathrm{i}^{|n|}\phi_{|n|} \\ 0 \\ 0 \end{pmatrix}, \quad \psi_{K',n} = C_n \begin{pmatrix} 0 \\ 0 \\ \mathrm{i}^{|n|}\phi_{|n|} \\ \mathrm{sgn}(n)\mathrm{i}^{|n|-1}\phi_{|n|-1} \end{pmatrix},$$

where $C_n = 1$ for $n = 0$ and $C_n = 1/\sqrt{2}$ for $n \neq 0$. Here the upper two terms correspond to the wave functions of A and B sites for the K valley and the lower two are those of the A and B sites for the K' valley. The two non-zero elements in $\psi_{K,n}(\psi_{K',n})$ correspond to the occupation of the sublattice A (upper term) and sublattice B (lower term). Here ϕ_n is the wave function for a particle with conventional parabolic dispersion in the n-th LL (see Chapter 2). It is clear that a specific feature of the Dirac fermion dispersion law is the *mixture* of the conventional LLs. This mixture is present only for $n \neq 0$. For $n = 0$ the electron in the valley K

---

[36]Unconventional integer quantum Hall effect in graphene, by V.P. Gusynin and S.G. Sarapov, *Phys. Rev. Lett.* **95**, 146801 (2005).

[37]Experimental observation of the quantum Hall effect and Berry's phase in graphene, by Y. Zhang, et al., *Nature* **438**, 201 (2005); Two-dimensional gas of massless Dirac fermions in graphene, by K.S. Novoselov et al., *Nature* **438**, 197 (2005); Half integer quantum Hall effect in high mobility single layer epitaxial graphene, by X. Wu et al., *Appl. Phys. Lett.* **95**, 223108 (2009).

[38]Quantum Hall effect in graphene, by Z. Jiang et al., *Solid State Commun.* **143**, 14 (2007).

[39]Hall conductivity of a two-dimensional graphite system, by Y. Zheng and T. Ando, *Phys. Rev. B* **65**, 245420 (2002).

or K' occupies only the sublattice A or B, respectively. For higher LLs the electron in each valley occupies both sublattices. The wave functions in the sublattices A and B are the wave functions of the conventional electrons with different LL indices. This property of the Dirac electrons in graphene strongly modifies the electron-electron interaction within a single Dirac fermion LL and influences the Dirac counterpart of the fractional quantum Hall effect (see Sect. 2.3.3) in graphene. From the analysis above, Apalkov and Chakraborty[40] predicted the existence of the Dirac form of the fractional quantized Hall effect in graphene. The FQHE in graphene has turned out to be very different from that in GaAs-based 2DEG. Electrons in graphene are more planar than in conventional semiconductor systems. The inter-electron interactions in graphene are stronger due to the absence of substrate screening. Stronger interactions results in a larger energy gap. The effect was subsequently confirmed in experiments[41], where substrate-induced perturbations were found to be a major hindrance in observing the effect[42]. A detailed analysis of the FQHE states in graphene is also available in the literature[43].

5.4 Bilayer graphene

Bilayer graphene can be created for example, by stacking two graphene layers in the same way as the stacking occurs in graphite, i.e., the Bernal stacking[44]. As demonstrated in Fig. 5.1, each graphene layer has a hexagonal (honeycomb) carbon lattice which is composed of two periodic sublattices A and B. In other words, there are two inequivalent lattice sites with atoms A and B each with unit cell of the periodic lattice. The two sublattices are displaced from each other along an edge of the hexagons by a distance of $a_0 = 1.42$Å. In a graphene bilayer, there are four

---

[40]Fractional quantum Hall states of Dirac electrons in graphene, by V.M. Apalkov and T. Chakraborty, *Phys. Rev. Lett.* **97**, 126801 (2006).

[41]See for example, Measurement of the $\nu = \frac{1}{3}$ fractional quantum Hall energy gap in suspended graphene, by F. Ghahari, et al., *Phys. Rev. Lett.* **106**, 046801 (2011).

[42]Fractional quantum Hall effect in suspended graphene probed with two-terminal measurements, by I. Skachko, et al., *Phil. Trans. R. Soc. A* **368**, 5403 (2010).

[43]Aspects of the fractional quantum Hall effect in graphene, by T. Chakraborty and V. Apalkov, in *Physics of Graphene*, edited by H. Aoki and M.S. Dresselhaus (Springer, New York 2014).

[44]Landau level degeneracy and quantum Hall effect in a graphite bilayer, by E. McCann and V.I. Fal'ko, *Phys. Rev. Lett.* **96**, 086805 (2006).

inequivalent sites in each unit cell, with atoms A and B at the top and A' and B' at the bottom. In Bernal stacking, the two layers are arranged such that the A sublattice is exactly on top of the sublattice B' with a vertical separation[45] of $b_0 = 3.4$Å [Figs. 5.6 (a) and 5.6(b)]. The system can be described by a tight-binding model[46] characterized by the coupling parameters $\gamma_0 = 3.16$ eV between the atoms A and B or A' and B' (intralayer coupling), $\gamma_1 = 0.39$ eV between A and B' (the direct interlayer coupling), and $\gamma_3 = 0.315$ eV between A' and B, A and A' or B and B' (the indirect interlayer coupling).

In the momentum space, the graphene bilayer has the same hexagonal Brillouin zone as that of a graphene monolayer (see Fig. 5.2 in Sec. 5.1.1). Its physical properties are mainly determined by the energy spectrum and the wave function near the two inequivalent corners of the Brillouin zone, K and K', where the π^* conduction band and the π valence band meet at the Fermi surface (see Sec. 5.1.1). Due to strong interlayer coupling (the π orbit overlap) both the conduction band and the valence band in a bilayer split by an energy of ~ 0.4 eV near the K and K' valleys[47]. Since this energy splitting is larger than the energy range we are interested in from the bottom of the energy band, we consider only the upper valence band and the lower conduction band. It should be noted that the graphene bilayer system cannot be treated as two independent graphene monolayers with the interlayer coupling as a perturbation because of the strong interlayer overlap of the π orbits.

Just as in monolayer graphene, in the effective-m. ass approximation[18] the electrons in the K valley are described by a Hamiltonian with

---

[45]Interplanar binding and lattice relaxation in a graphite dilayer, by S.B. Trickey, F. Müller-Plathe, G.H.F. Diercksen, and J. C. Boettger, *Phys. Rev. B* **45**, 4460 (1992).

[46]The electronic properties of bilayer graphene, by E. McCann and M. Koshino, *Rep. Prog. Phys.* **76**, 056503 (2013).

[47]Interlayer interactions in two-dimensional systems: Second-order effects causing ABAB stacking of layers in graphite, by K. Yoshizawa, T. Kato, and T. Yamabe, *J. Chem. Phys.* **105**, 2099 (1996).

[48]Electronic properties of monolayer and bilayer graphene, by E. McCann, in *Graphene Nanoelectronics*, edited by H. Raza (Springer, Heidelberg 2012).

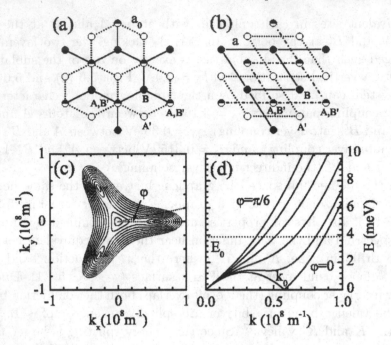

FIGURE 5.6
(a) The hexagonal lattice of the top graphene layer (solid line) and the bottom layer (dotted line). (b) Periodic sublattices A and B' (dash-dotted lines), B (solid lines), and A' (dotted). In (a) and (b), the atoms of the sublattice $A(B'), B$, and A' are denoted by semisolid, solid, and open circles, respectively. (c) The contour lines of the energy in the $k_x - k_y$ plane near the K point. The corresponding energies, starting with the innermost curve, are $0.1E_0$ to E_0 with an increment of $0.1E_0$. (d) The energy spectrum for equally separated φ from 0 to $\pi/6$ (curves with increasing energy).

a mixture of the linear and the quadratic terms[49] of k:

$$\mathcal{H}_K = \frac{\hbar^2}{2m^*} \begin{pmatrix} 0 & k_-^2 \\ k_+^2 & 0 \end{pmatrix} - \frac{\hbar^2 k_0}{2m^*} \begin{pmatrix} 0 & k_+ \\ k_- & 0 \end{pmatrix},$$

with $k_\pm = k_x \pm i k_y$ and $\mathbf{k} = (k_x, k_y)$ being measured from the K point. The effective mass of the quadratic term is $m^* = 2\hbar^2\gamma_1 / (3a_0\gamma_0)^2 \approx$

---

[49]Coulomb screening and collective excitations in a graphene bilayer, by X.-F. Wang and T. Chakraborty, *Phys. Rev. B* **75**, 041404(R) (2007).

$0.033m_0$ with m_0 being the free electron mass and the 'light' velocity (drawing from the concept in monolayer graphene) of the linear term is $v_0 = \hbar k_0/2m^* = 3a_0\gamma_3/2\hbar \approx 10^5$ m/s with $k_0 \approx 10^8/\sqrt{3}$ m^{-1}.

The eigenfunction of the above Hamiltonian is $\Psi_k^\lambda(\mathbf{r}) = \frac{e^{i\mathbf{k}\cdot\mathbf{r}}}{\sqrt{2}}\begin{pmatrix} e^{i\phi_k} \\ \lambda \end{pmatrix}$ with the energy $E_k^\lambda = \lambda\hbar^2 k\sqrt{k^2 - 2k_0 k\cos 3\varphi + k_0^2}/2m^*$ and the pseudospin angle $\lambda\phi_k$. Here $\varphi = \arg(k_+), \phi = \arg(ke^{-2i\varphi} - k_0 e^{i\varphi})$ with $\arg(z)$ being the argument θ of a complex $z = |z|e^{i\theta}$ and $\lambda = 1(-1)$ for the conduction (valence) band.

For $k \gg k_0$ the electron states are chiral (as elaborated in Sec. 5.1.1) with $\phi = -2\varphi$ and have an approximately isotropic *parabolic* energy dispersion $E_k^\lambda = \lambda\hbar^2 k^2/2m^*$. Near $k = k_0$, the energy dispersion becomes highly anisotropic [Figs. 5.6 (c) and 5.6 (d)]. The corresponding characteristic energy is $E_0 = \hbar^2 k_0^2/2m^* = 3.9$ meV. The electrons are now described as *massive chiral fermions*[50]. At $E = 0$, where the Fermi energy is located for undoped bilayer graphene, there are four contact points between the conduction and valence bands: one at $k = 0$, the center of the valley, and the three satellites at $k = k_0$ in the directions of $\psi = 0, 2\pi/3$, and $4\pi/3$. They can be treated as four Dirac points because the electronic states near each point have a linear energy dispersion and have the same chirality as those near a Dirac point in monolayer graphene. However, compared to the monolayer graphene, the energy dispersion here is anisotropic and the 'light' velocity is about 10 times slower. As shown in Figs. 5.6 (c) and 5.6 (d), there is an energy pocket with a depth of about $E_0/4$ at each Dirac point. These characteristics of the bilayer graphene make this system very different from monolayer graphene.

5.4.1 Bilayer graphene Landau levels

In an applied perpendicular magnetic field, the LL energies of the massive Dirac fermions in bilayer graphene takes the form, $\mathcal{E}_n = \pm\hbar\omega_c\sqrt{n(n-1)}$, where $n = 0, 1, 2, \ldots$ is the Landau orbit index in each layer, and $\omega_c = eB/m^*$ is the cyclotron frequency and m^* is the effective mass of the particles. The LL sequence is linear in B, just as for conventional electron gas (Chapter 2). However, there are zero-energy ($\mathcal{E} = 0$) LLs at $n = 0, 1$ that are independent of the magnetic field. For

---

[50]Berry phase and pseudospin winding number in bilayer graphene, by C.-H. Park and N. Marzari, *Phys. Rev. B* **84**, 205440 (2011).

$n > 1$, the LLs are four-fold degenerate (accounting for spin and valley), similar to those in monolayer graphene. However, the $n = 0, 1$ LLs of bilayer graphene are eightfold degenerate zero-energy states[51] arising from the fact that the wavefunctions for both these states are localized on either the top or bottom layer. This is reflected in the integer quantum Hall effect. The Hall resistance for bilayer graphene would display a series of quantized plateaus occurring at integer values of $4e^2/h$ that is essentially the same as the conventional case. However, there is a step of size $8e^2/h$ in the Hall resistance across zero density due to the eightfold degeneracy unique to bilayer graphene[52].

5.4.2 Novel fractional quantum Hall effects

An important characteristic of bilayer graphene is that it is a semiconductor with a tunable bandgap between the valence and conduction bands[53]. This property modifies the LL spectrum and influences the role of long-range Coulomb interactions[54]. For the AA stacking of bilayer graphene (Fig. 5.7) in a perpendicular magnetic field, the interlayer coupling occurs between the LLs of the two layers with the same LL indices. This coupling changes the energies of the LLs of the monolayers, but does not alter the wave functions of the layers. Therefore, the electron-electron interactions have the same characteristics as that of a monolayer graphene.

In a magnetic field, the electronic properties of the graphene bilayer can be controlled by the inter-layer bias voltage applied to the two layers. The technical details for biased bilayer graphene are briefly presented in Box. 5.2[55], and can also be found in Footnote 5.

---

[51]Electronic properties and the quantum Hall effect in bilayer graphene, by V.I. Fal'ko, *Phil. Tran. R. Soc. A* **366**, 205 (2008).

[52]Symmetry breaking in the zero-energy LL in bilayer graphene, by Y. Zhao, et al., *Phys. Rev. Lett.* **104**, 066801 (2010).

[53]Landau levels and oscillator strength in a biased bilayer of graphene, by J.M. Pereira, Jr. et al., *Phys. Rev. B* **76**, 115419 (2007).

[54]Long-Range Coulomb Interaction in Bilayer Graphene, by D.S.L. Abergel and T. Chakraborty, *Phys. Rev. Lett.* **102**, 056807 (2009).

[55]A detailed description of interacting electrons in bilayer graphene in the quantum Hall regime are available in Traits and characteristics of interacting Dirac fermions in monolayer and bilayer graphene, by T. Chakraborty and V.M. Apalkov, *Solid State Commun.* **175-176**, 123 (2013).

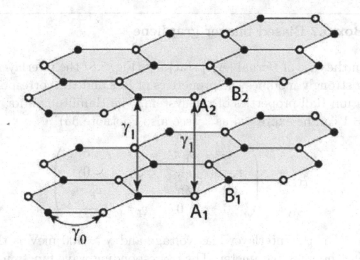

FIGURE 5.7

AA stacking of bilayer graphene. The intra-layer and inter-layer coupling parameters are denoted by γ_0 and γ_1, respectively.

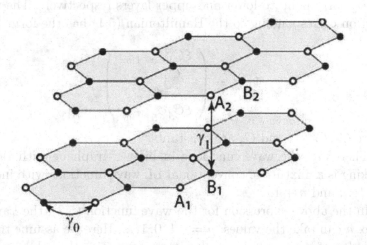

FIGURE 5.8

Bernal stacking of bilayer graphene: The intra-layer and inter-layer coupling parameters are denoted by γ_0 and γ_1, respectively.

Box 5.2 Biased bilayer graphene

In the case of Bernal (AB) stacking (Fig. 5.8) the interlayer coupling strongly modifies the properties of LLs and the corresponding quantum Hall properties of the system. The Hamiltonian for valley $\xi = \pm 1$ can be expressed as[56] (see also Footnote 55)

$$
\mathcal{H}_\xi = \xi \begin{pmatrix} \frac{U}{2} & v_F \pi_- & 0 & 0 \\ v_F \pi_+ & \frac{U}{2} & \xi\gamma_1 & 0 \\ 0 & \xi\gamma_1 & -\frac{U}{2} & v_F \pi_- \\ 0 & 0 & v_F \pi_+ & -\frac{U}{2} \end{pmatrix},
\tag{5.1}
$$

where U is the interlayer bias voltage and $\gamma \approx 400$ meV is the interlayer hopping parameter. The corresponding wave function is described by a four-component spinor $\left(\psi_{A_1}, \psi_{B_1}, \psi_{B_2}, \psi_{A_2}\right)^{\mathrm{T}}$ for valley K and $\left(\psi_{B_2}, \psi_{A_2}, \psi_{A_1}, \psi_{B_1}\right)^{\mathrm{T}}$ (where the superscript T indicates the transpose of a vector) for valley K'. The subindices A_1, B_1 and A_2, B_2 correspond to lower and upper layers respectively. The wave function corresponding to the Hamiltonian (5.1) has the form

$$
\Psi_{n,m}^{(bi)} = \begin{pmatrix} \xi C_1 \phi_{n-1,m} \\ C_2 \phi_{n,m} \\ C_3 \phi_{n,m} \\ \xi C_4 \phi_{n+1,m} \end{pmatrix}
\tag{5.2}
$$

where C_1, C_2, C_3 and C_4 are constants.[57]

Therefore, the wave functions in bilayer graphene with Bernal stacking is a mixture of conventional LL wave functions with indices $n-1, n$, and $n+1$.

In the above expression for the wave function (5.2), the Landau index n can take the values: $n = -1, 0, 1, ...$ Here we assume that if the index of the Landau wave function is negative then the function is identically zero, i.e., $\phi_{-2,m} \equiv 0$ and $\phi_{-1,m} \equiv 0$. In this case, for $n = -1$ the wave function (5.2) is just $\Psi_{-1,m}^{(bi)} = (0,0,0,\phi_{0,m})$, i.e., the coefficients C_1, C_2, C_3 are equal to zero. There is only one energy level corresponding to $n = -1$. For $n = 0$, the wave function (5.2) has a zero coefficient C_1, which results in three energy levels corresponding to $n = 0$. For other values of n, i.e., for $n > 0$, there are four eigenvalues of the Hamiltonian (5.1), corresponding to four LLs in a bilayer for a given valley $\xi = \pm 1$.

The eigenvalue equation determining these LLs have the form (see Footnote 53):

$$\left[(\mathcal{E} + \xi u)^2 - 2n\right]\left[(\mathcal{E} - \xi u)^2 - 2(n+1)\right] = \tilde{\gamma}_1^2 \left[\mathcal{E}^2 - u^2\right], \quad (5.3)$$

where $\mathcal{E}$ is the energy of the LL in units of $\mathcal{E}_B$ ($\mathcal{E}_B = \hbar v_F/\ell_0$), $u = U/(2\mathcal{E}_B)$, and $\tilde{\gamma}_1 = \gamma_1/\mathcal{E}_B$.

We introduce the following labeling scheme for the LLs determined by the above eigenvalue equation [Eq. (5.3)]. The four LLs correspond to two valence levels that have negative energies, and two conduction levels, that have positive energies. Then the four LLs of bilayer graphene for a given value of n and a given valley ξ can be labeled as $n_i^{(\xi)}$, where $i = -2, -1, 1, 2$ is the label of the LL in the ascending order. Here negative and positive values of i correspond to the valence and conduction levels, respectively.

The LLs of different valleys are related through the equation: $\mathcal{E} = -\mathcal{E}\left(n_i^{(\xi)}\right) = -\mathcal{E}\left(n_{-i}^{(-\xi)}\right)$. Although for $n = 0$ there are only three LLs and for $n = -1$ there is only one LL, it is convenient to include the $n = -1$ LL into the set of $n = 0$ LLs and label them as $0_i^{(\xi)}$, where $i = -2, -1, 1, 2$.

At the zero bias voltage, the LLs become two-fold valley and two-fold spin degenerate and are expressed as

$$\mathcal{E} = \pm\sqrt{2n + 1 + \frac{\tilde{\gamma}_1^2}{2} \pm \frac{1}{2}\sqrt{(2 + \tilde{\gamma}_1^2)^2 + 8n\tilde{\gamma}_1^2}}. \quad (5.4)$$

In what follows, we only consider the sets of LLs of bilayer graphene with $n = 0, 1$ only. The wave functions of these LLs are mixtures of the conventional LL wave functions with indices 0,1, and 2.

There are two special LLs of bilayer graphene. For $n = -1$ there are two solutions (one for the valley K and one K') of Eq. (5.3) with energies $\mathcal{E} = -\xi u$.

The corresponding wave function

$$\Psi^{(bi)}_{0^{(+)}_1,m} = \Psi^{(bi)}_{0^{(-)}_{-1},m} = \begin{pmatrix} 0 \\ 0 \\ 0 \\ \phi_{0,m} \end{pmatrix}, \tag{5.5}$$

is clearly determined only by the $n = 0$ conventional LL wave function. Therefore, the properties of interacting electrons, in particular the fractional quantum Hall effect (FQHE) of these LLs are exactly the same as those for the zeroth conventional LL.

For $n = 0$ and for small values of u there is another solution of Eq. (5.3) with almost zero energy, $\mathcal{E} \approx 0$. The wave function of this LL has the form

$$\Psi^{(bi)}_{0^{(+)}_{-1},m} = \Psi^{(bi)}_{0^{(-)}_1,m} = \frac{1}{\sqrt{\tilde{\gamma}_1^2 + 2}} \begin{pmatrix} 0 \\ \sqrt{2}\phi_{0,m} \\ 0 \\ \tilde{\gamma}_1\phi_{1,m} \end{pmatrix}$$

$$= \frac{1}{\sqrt{\gamma_1^2 + 2\mathcal{E}_B^2}} \begin{pmatrix} 0 \\ \sqrt{2}\mathcal{E}_B\phi_{0,m} \\ 0 \\ \gamma_1\phi_{1,m} \end{pmatrix}. \tag{5.6}$$

For a small magnetic field, $\mathcal{E}_B \ll \gamma_1$, the wave function becomes $(\psi_{1,m},0,0,0)^T$ and the LL is then identical to the $n = 1$ conventional LL. In a large magnetic field $\mathcal{E}_B \gg \gamma_1$, the LL wave function becomes $(0,0,\psi_{0,m},0)^T$ and the bilayer LL has the same properties as that of the $n = 0$ conventional LL.

For all the LLs [except the levels described by Eq. (5.5)] in bilayer graphene the electron-electron interaction strength and the stability (i.e., the energy gap) of the FQHE states depend on the magnetic field B, the

[56]Asymmetry gap in the electronic band structure of bilayer graphene, by E. McCann, *Phys. Rev. B* **74**, 161403 (2006).

[57]The full expressions of these constants are available in Footnote 43.

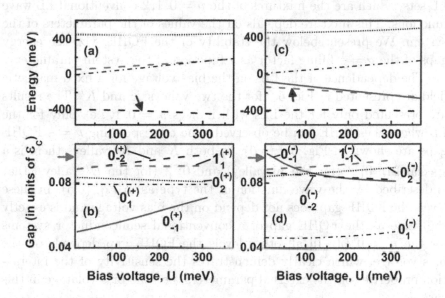

FIGURE 5.9

LLs of the bilayer graphene [panels (a) and (c)] versus the bias voltage
U. The energy gaps (in units of the Coulomb energy) of the $\frac{1}{3}$-FQHE in
corresponding LLs [panels (b) and (d)]. The numbers next to the lines
are the labels of the LLs. The same type of lines [in panels (a)-(b) and
panels (c)-(d)] correspond to the same LLs. Panels (a)-(b) correspond
to valley K, while panels (c)-(d) correspond to valley K'. The system is
characterized by $\gamma_1 = 400$ meV and a magnetic field of $B = 15$ Tesla.
The arrows in panels (a) and (c) show the LL with the strongest $\frac{1}{3}$-
FQHE. The arrows in (b) and (d) indicate the gap of $\frac{1}{3}$-FQHE in the
$n = 1$ LL of monolayer graphene.

bias voltage U, and the LL index. Therefore the interaction properties of a bilayer graphene can be controlled by the external parameters. This is certainly a different situation from the case of monolayer graphene where the interaction properties solely depend on the LL index. The stable FQHE states in a bilayer graphene are expected for the $n = 0, 1$ LL sets, which are the mixtures of the $n = 0, 1, 2$ conventional LL wave functions. This mixture depends on the values of the parameters of the system. We present below the stability of the FQHE, viz. the energy gaps of the $\nu = \frac{1}{3}$ filling factor as a function of the system parameters.

The dependence of the LLs on the bias voltage for a fixed magnetic field are presented in Fig. 5.9 for the two valleys K and K'. The results are presented only for the LLs with indices $n = 0, 1$, i.e., only for the LL where the FQHE can be observed. The corresponding $\nu = \frac{1}{3}$ FQHE gaps are shown in Fig. 5.9 (b,d). In both K and K' valleys there is a special LL, $0_1^{(+)}$ (for the K valley) and $0_{-1}^{(-)}$ (for the K' valley), that is described by the wave function of the type as in Eq. (5.5). In these levels the FQHE gap does not depend on the bias voltage and is exactly the same as the FQHE gap of a conventional semiconductor systems for the $n = 0$ LL. In all other levels the FQHE gap depends on the bias voltage, which clearly demonstrates the sensitivity of the interaction properties on the external parameters, i.e., the bias volatage in this case.

From Fig. 5.9 it is quite clear that in each valley the bilayer graphene has four LLs with a strong $\frac{1}{3}$-FQHE for all values of the parameters of the system. These are $0_{-2}^{(+)}, 0_1^{(+)}, 0_2^{(+)}, 1_1^{(+)}$ (valley K) and $0_{-2}^{(-)}, 0_{-1}^{(-)}, 0_2^{(-)}, 1_{-1}^{(-)}$ (valley K'). Therefore, for a given valley there are three LLs with $n = 0$ and one LL with $n = 1$, which show stable FQHE. The gaps of the corresponding FQHE states are usually between the gaps in the $n = 0$ and $n = 1$ FQHE state in monolayer graphene. The value of the $\frac{1}{3}$-FQHE gap in the $n = 1$ LL of monolayer graphene is shown by arrows in Fig. 5.9 (b), (d). In the LL $0_{-2}^{(+)}$ the FQHE state becomes more stable than the corresponding state in monolayer graphene.

The LL with the label $0_{-1}^{(+)}$ shows a strong dependence of the interaction properties on the parameters of the system. With increasing bias voltage the FQHE gap and correspondingly its stability increases rapidly. At a fixed filling factor of bilayer graphene, this type of behavior can result in an unique transition from a non-FQHE state at a small bias voltage (for example) to a FQHE state at large bias voltages. Precisely

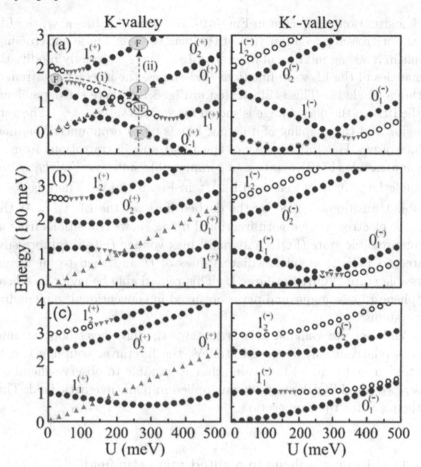

FIGURE 5.10

A few lowest LLs of the conduction band as a function of the bias potential U, for different values of the inter-layer coupling: (a) $\gamma_1 = 30$ meV, (b) $\gamma_1 = 150$ meV and (c) $\gamma_1 = 300$ meV and a magnetic field of 15 Tesla. The numbers next to the curves denote the corresponding LLs. The LLs where the FQHE can be observed are drawn as • and ▲ points. The ▲ points correspond to the LLs where the FQHE states are identical to that of a monolayer graphene or a conventional semiconductor system. The ▼ points represent LLs with weak FQHE. The open dots correspond to LLs where the FQHE is absent. In (a), the dashed lines labeled by '(i)' illustrates the transition between FQHE (symbol 'F') and no FQHE (symbol 'NF') states under a constant gate voltage and variable bias potential, and '(ii)' under a constant bias potential but variable gate voltage. Left and right columns correspond to the K and K' valleys, respectively.

such a situation is depicted in Fig. 5.10. For a small value of γ_1, the LLs show anticrossings as a function of the bias voltage. These anticrossings result in a strong mixture of different LLs, that can strongly modify the properties of the LL wave functions and change the interaction strength within a single LL. This is illustrated in Fig. 5.10, where the dependence of the LLs on the bias voltage is shown for various values of γ_1. The anticrossing and the coupling of different LLs is much pronounced for small values of γ_1. The anticrossings of the LLs result in transitions from an incompressible (FQHE) state to a compressible state (no FQHE) within a single Landau level [see the LL $1_2^{(+)}$ in Fig. 5.10 (a)]. There is also a double transition, marked by the dashed lines, at the LL $1_1^{(+)}$. At this LL, the electron system with fractional filling shows transitions from an incompressible state (FQHE) at small bias voltage U to a compressible state (no FQHE) at intermediate values of U and then to an incompressible state (FQHE) at large U. This remarkable behavior in bilayer graphene LLs is unique and never occurred in conventional semiconductor systems[58].

While for the available bilayer systems the inter-layer hopping integral is relatively large, $\gamma_1 \approx 400$ meV, the interlayer coupling can be controlled and reduced to a value that is suitable to observe the above described novel FQHE states by an applied in-plane magnetic field. That is the topic for the next section.

5.4.3   Bilayer graphene in a tilted magnetic field

A tilted magnetic field, applied to a quasi-two-dimensional electron system, can modify the electron dynamics and correspondingly the electron-electron interaction strength. In monolayer graphene, due to its purely two-dimensional nature, the component of the magnetic field parallel to the monolayer does not influence the electron's spatial dynamics, although it can alter the electron spin dynamics which is sensitive to the total magnetic field (see Sec. 2.7).

Bilayer graphene is a quasi-two-dimensional system where the electron dynamics is sensitive to both perpendicular and in-plane

---

[58]Controllable driven phase transitions in fractional quantum Hall states in bilayer graphene, by V.M. Apalkov and T. Chakraborty, *Phys. Rev. Lett.* **105**, 036801 (2010).

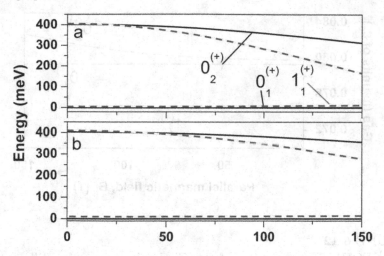

FIGURE 5.11
The LLs of bilayer graphene in a tilted magnetic field and zero bias voltage as a function of the parallel component of the magnetic field. The perpendicular component of the field is (a) 1 Tesla and (b) 2 Tesla. The labels next to the lines denote the corresponding LLs (only those that shows the FQHE are labeled). Only the LLs with positive energies are shown.

components of the magnetic field[59]. The LLs of a bilayer graphene in a tilted magnetic field are given by Eq. (5.4), in which the interlayer coupling $\tilde{\gamma}_1$ is replaced by $\tilde{\gamma}_{1,n} = \gamma_{1,n}/\mathcal{E}_B$. Here $\mathcal{E}_B$ is evaluated for the perpendicular component of the tilted magnetic field, $\gamma_{1,n} = \gamma_1 \kappa_n(\mu)$, where $\kappa_n(\mu) = \int dx\, \psi_k(x)\psi_{k-\beta}(x)$, $\beta = eB_\parallel d/\hbar$, d is the interlayer separation, and $\mu = \beta\ell_0$. The magnetic length $\ell_0 = \sqrt{\hbar/eB_\perp}$ is now defined in terms of the perpendicular component of the magnetic field. The effect of the in-plane component of the magnetic field on the electron dynamics in a bilayer graphene is the reduction of the interlayer coupling, $\gamma_{1,n} < \gamma_1$. The dependence of the LLs on the parallel components of the magnetic field, $B_\parallel$, is shown for a few lowest LLs of the bilayer graphene.

Increasing the parallel component of the field the energies of the LLs are reduced, which is consistent with the reduction of the interlayer

---

[59]Landau level spectrum for bilayer graphene in a tilted magnetic field, by Y.-H. Hyun, et al., *J. Phys.: Condens. Matter* **24**, 045501 (2012).

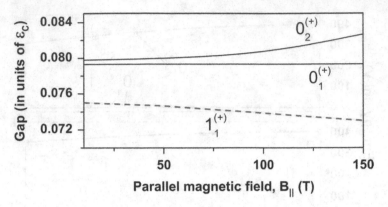

FIGURE 5.12

The $\frac{1}{3}$-FQHE gaps (in units of the Coulomb energy) for different LLs of bilayer graphene versus the parallel component of the tilted magnetic field. The perpendicular component of the magnetic field is 1 T. The labels next to the lines denote the corresponding LLs (Fig. 5.11). The bias voltage is zero in this case.

coupling $\gamma_{1,n}$ with increasing $B_\parallel$. The dependence of the LLs on $B_\parallel$ becomes weaker with increasing perpendicular field [Fig. 5.11 (a, b)]. Therefore, the effect of an in-plane field on the LLs can be observed only for a small perpendicular magnetic field ($B_\perp \approx 1$), and a large parallel field ($B_\parallel \geq 50$ Tesla). We should hasten to add that although the perpendicular component of the field is rather small, in a conventional semiconductor system the FQHE has been reported in a magnetic field of $B < 3$ Tesla[60].

The properties of interacting electrons in bilayer LLs also depend on the in-plane component of the magnetic field. This dependence is visible only for small perpendicular component of the magnetic field, i.e., $B_\perp \approx 1$ Tesla. The $\frac{1}{3}$-FQHE gap as a function of the in-plane component of the field is shown in Fig. 5.11, for different LLs. For $B_\perp \approx 1$ Tesla, only three LLs (with positive energies) support the FQHE states. One LL $0_1^{(+)}$, the wave function of which has the form of (5.5) and depends only on the perpendicular component of the magnetic field, does not show any dependence on the in-plane component of the magnetic field.

---

[60]Spin polarization of two-dimensional electrons in different fractional states and around filling factor $\nu = 1$, by I. V. Kukushkin, K. v. Klitzing, and K. Eberl, *Phys. Rev. B* **55**, 10607 (1997).

The interaction strength in the LLs $0_2^{(+)}$ and $1_1^{(+)}$ depends weakly on $B_\parallel$ (Fig. 5.11). The interaction strength increases with $B_\parallel$ for the LL $0_2^{(+)}$ and decreases for the LL $1_1^{(+)}$. Therefore the parallel component of the magnetic field can in fact, *enhance* the electron-electron interaction strength for some LLs in a bilayer graphene. Footnote 55 contains a more detailed account of the tilted-field effects on electrons in a bilayer graphene.

5.4.4 Marvels of interacting electrons in graphene

Interacting electrons in graphene can host several fascinating phenomena. Here we list only a few[61] without elaborating on any further details. Bilayer graphene is predicted to support the so-called Pfaffian state[62] that was originally proposed for the $\nu = \frac{5}{2}$ fractional quantum Hall states in conventional two-dimensional electron systems. Valley polarized current is predicted to be generated in bilayer graphene by subjecting it to intense terahertz frequency light in the presence of a transverse electric field[63]. Two parallel graphene layers separated by a thin insulator can induce interlayer-correlated fractional quantum Hall states[64]. Novel interaction-induced transitions in the fractional quantum Hall states that can be tuned entirely by the applied bias voltage are predicted in trilayer graphene[65]. A rich profusion of interaction-induced effects in twisted bilayer or trilayer graphene has been observed in recent years[66].

---

[61]Footnotes 5-7, 11 contain more detailed descriptions of the physical properties of graphene.

[62]Stable Pfaffian state in bilayer graphene, by V.M. Apalkov and T. Chakraborty, *Phys. Rev. Lett.* **107**, 186803 (2011).

[63]Generation of valley polarized current in bilayer graphene, by D.S.L. Abergel and T. Chakraborty, *Appl. Phys. Lett.* **95**, 062107 (2009).

[64]Interlayer fractional quantum Hall effect in a coupled graphene double layer, by X. Liu, et al., *Nat. Phys.* **15**, 893 (2019).

[65]Electrically tunable charge and spin transitions in Landau levels of interacting Dirac fermions in trilayer graphene, by V.M. Apalkov and T. Chakraborty, *Phys. Rev. B* **86**, 035401 (2012).

[66]Elecric field tunable superconductivity in alternating-twist magic-angle trilayer graphene, by Z. Hao, et al., *Science* **371**, 1133 (2021), and references therein.

5.5   Graphene nanostructures

In spite of the many spectacular properties of graphene, described in previous sections, there is a major challenge in utilizing graphene as a building block for nanoelectronics: graphene lacks a band gap. The absence of which makes graphene unsuitable to create a field effect transistor. One possible remedy of this problem is to make graphene nanostructures, such as quantum dots or quantum rings. As described in Chapter 3, in conventional semiconductors the charge carriers can be confined by potential barriers that are created by electrostatic gates. This approach allows for control of individual electrons in the system.

5.5.1   Quantum dots

Confinement of Dirac fermions in graphene would be of major importance for making nanoelectronic devices from graphene to explore the optical and transport spectroscopy, and exploit the spin and valley degrees of freedom. However, it is not possible to confine chiral Dirac fermions, not least the massless ones (think of the neutrinos). Making conventional quantum dots is not possible with graphene. In this case, we need to focus on *trapping* the electrons. This means to keep the electrons remain within a finite spatial region for a considerable amount of time.

The problem of trapping electrons was investigated for quantum dots with smooth[67] and sharp[68] boundaries. The trapping potential for Dirac fermions in graphene is generally produced by the transverse momentum[69]. In a parabolic quantum dot, the transverse potential is related to the electron angular momentum: The larger the angular momentum the more efficient is the trapping potential.

The trapping time is also strongly influenced by the sharpness of the QD boundaries, i.e., the sharpness of the confinement potential. The

---

[67]Fock-Darwin states of Dirac electrons in graphene-based artificial atoms, by Hong-Yi Chen, V. Apalkov and T. Chakraborty, *Phys. Rev. Lett.* **98**, 186803 (2007).

[68]Quasibound states of quantum dots in single and bilayer graphene, by A. Matulis and F.M. Peeters, *Phys. Rev. B* **77**, 115423 (2008).

[69]Quantum dots in graphene, by P.G. Silvestrov and K.B. Efetov, *Phys. Rev. Lett.* **98**, 016802 (2007), Selective transmission of Dirac electrons and ballistic magnetoresistance of n-p junctions in graphene, by V.M. Cheianov and V.I. Fal'ko, *Phys. Rev. B* **74**, 041403 (2006).

sharpness of the boundary of a QD determines the width of the trapping potential. This means that the most efficient trapping is possible with a smooth confinement potential and for electronic states with large angular momentum. The trapping time of an electron in a QD has exponential dependence on the angular momentum and the slope of the confinement potential. While the most efficient trapping is possible with a smooth potential, some trapping can also be expected in a confinement potential with sharp boundaries. Here the trapping time has no exponential dependence but instead have a power dependence on the parameters of the confinement potential.

There are other possible ways, as proposed in the literature to confine Dirac electrons, namely, cutting samples into different shapes. Nanostructures, such as nanoribbons[70] or single-electron transistors[71] have been created this way. Such a QD has been utilized to create a graphene charge detector[72]. Confinement by magnetic barriers[73] or by a combination of electric and magnetic fields[74] have also been proposed.

The dipole-allowed optical transitions, just as for conventional QDs described in Chapter 3, were studied for parabolic graphene QDs (see Footnote 66). In the absence of the confinement potential, there is just a single transition (between the lowest two LLs), but there are additional transitions in the presence of the parabolic confinement. At high magnetic fields, it has been proposed that, the optical absorption spectra can be useful to determine the band parameters in graphene.

Pristine bilayer graphene has a gapless spectrum which is parabolic at low energies near the two inequivalent points in the Brillouin zone (K and K'). In a perpendicular electric field, the spectrum displays a gap that can be tuned by varying the bias voltage. A parabolic QD can be realized by biasing nanostructures gates on the bilayer graphene. In contrast to the conventional semiconductor QDs, here the ground

---

[70]Energy band-gap engineering of graphene nanoribbons, by M.Y. Han, et al., *Phys. Rev. Lett.* **98**, 206805 (2007).

[71]Tunable Coulomb blockade in nanostructured graphene, by C. Stampfer, et al., *Appl. Phys. Lett.* **92**, 012102 (2008).

[72]Charge detection in graphene quantum dots, by J. Güttinger, et al., *Appl. Phys. Lett.* **93**, 212102 (2008).

[73]Magnetic confinement of massless Dirac fermions in graphene, by A. De Martino, et al., *Phys. Rev. Lett.* **98**, 066802 (2007).

[74]Tunable quantum dots in monolayer graphene, by G. Giavaras and F. Nori, *Phys. Rev. B* **85**, 165446 (2012).

state energy of the two-electron spectrum exhibits a valley transition in addition to the usual spin singlet-triplet transition[75].

5.5.2 Quantum rings

Theoretical studies of graphene quantum ring (QR) were initiated by Recher et al.[76], who used the Dirac model to analyze the electronic properties of Aharonov-Bohm (AB) rings made out of graphene (Fig. 5.13). They demonstrated that the ring confinement and the external magnetic field have the combined effect to break the valley degeneracy in graphene. If confirmed, it will open up the intriguing possibility to control valley polarization for device applications. The electron-electron interaction effects were however ignored in this work, thereby reducing the practicality of this idea. The important and interesting question of the interplay between valley polarization and the Coulomb interaction was first addressed by Abergel et al.[77]. These authors found that the interaction causes drastic changes in the nature of the ground state. The Coulomb interaction has a strong influence on the energy spectrum, the persistent current, and the optical-absorption spectrum of a graphene ring.

The ring confinement potential for the QR is defined by the infinite mass boundary conditions[78] in which case the massless Dirac Hamiltonian is coupled with a mass potential that is zero inside the ring (i.e., $V(r) = 0$) but is infinite elsewhere. The Coulomb interaction does not alter the valley state of the electron, and conserves the total angular momentum between the initial and final states. The exact diagonalization scheme was then applied to determine the energy and eigenstates of the interacting system. The persistent current j was then obtained by taking the derivative of the ground state energy E_0 of the few electron system with respect to the flux as $j(\Phi) = \frac{\partial}{\partial \Phi} E_0$.

---

[75]Electron-electron interactions in bilayer graphene quantum dots, by M. Zarenia et al., *Phys. Rev. B* **88**, 245432 (2013).

[76]Aharonov-Bohm effect and broken valley degeneracy in graphene rings, by P. Recher, et al., *Phys. Rev. B* **76**, 235404 (2007).

[77]Interplay between valley polarization and electron-electron interaction in a graphene ring, by D.S.L. Abergel, V.M. Apalkov and T. Chakraborty, *Phys. Rev. B* **78**, 193405 (2008).

[78]Neutrino billiards: time-reversal symmetry-breaking without magnetic fields, by M. V. Berry and R. J. Mondragon, *Proc. R. Soc. Lond. A* **412**, 53 (1987); Symmetry of boundary conditions of the Dirac equation for electrons in carbon nanotubes, by E. McCann and V.I. Fal'ko, *J. Phys.: Condens. Matter* **16**, 2371 (2004).

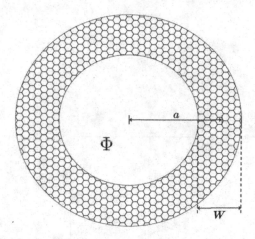

FIGURE 5.13
The geometry of a quantum ring. The magnetic field is perpendicular to the graphene plane.

The calculated intensity of the optical absorption is proportional to the area of the dots in the lowest panels of Fig. 5.14, Fig. 5.15 (a) and Fig. 5.15 (b) scale with this quantity. In all figures we show the absorption of unpolarized light where transitions which change the angular momentum quantum number by ± 1 are permitted, as long as the valley index remains unchanged. Where the initial state of a transition is degenerate, we take the average of the intensity of all possible pairs of initial and final states.

In Fig. 5.14 we show the energy spectrum, persistent current and optical absorption for a single electron in the graphene ring with $R/W = 10$. Lifting of the valley degeneracy causes the step in the persistent current at $\phi = \Phi/\Phi_0 = 0.5$. For $0 < \phi < 0.5$ the ground state consists of one electron in the $m = -\frac{1}{2}$, $\tau = -1$ state whereas for $0.5 < \phi < 1$ the valley index is $\tau = +1$. For $\phi \gtrsim 0$, transitions to the lowest-lying states $m = +\frac{1}{2}$, $\tau = +1$ and $m = -\frac{1}{2}$, $\tau = -1$ are not allowed since the optical absorption cannot mix valleys. Here m is the orbital angular momentum and τ denotes the Pauli matrices in the valley space.

For two non-interacting electrons, the ground state consists of a pair of electrons with anti-parallel spins occupying the same single-particle states as in the single-electron system [Fig. 5.15 (a)]. The persistent current reflects the similarity between the ground states of the single-particle and $N = 2$ non-interacting system, and since there are now

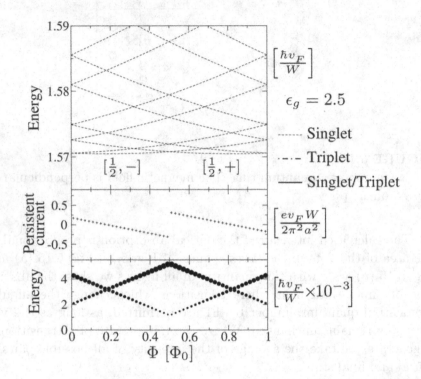

FIGURE 5.14

Energy spectrum (top pane), persistent current (middle pane) and optical absorption of unpolarized light (lower pane) as a function of Φ/Φ_0 for a single electron. The area of the points in the absorption plots represent the intensity of the peak in arbitrary units. In the plots, $W = 10$ nm and $\frac{R}{W} = 10$, and $\epsilon_g = 2.5$.

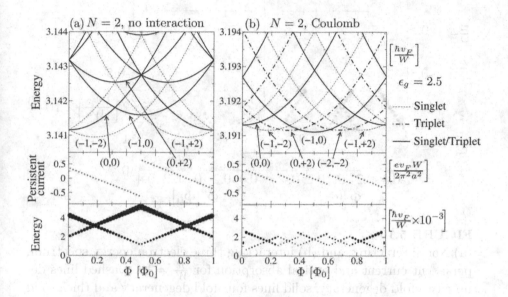

FIGURE 5.15

Energy spectrum (top pane), persistent current (middle pane) and optical absorption of unpolarized light (lower pane) as a function of Φ/Φ_0 for (a) two non-interacting electrons, and (b) two electrons with the Coulomb interaction included. The area of the points in the absorption plots represent the intensity of the peak in arbitrary units. In the plots, $W = 10$ nm and $\frac{R}{W} = 10$, and $\epsilon_g = 2.5$.

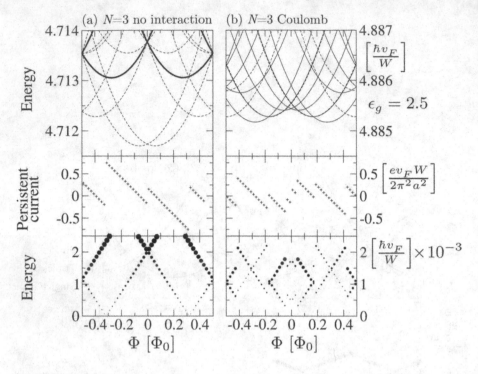

FIGURE 5.16
(a) Non-interacting, and (b) interacting three electron energy spectrum, persistent current and optical absorption for $\frac{R}{W} = 10$. Dashed lines denote two-fold degeneracy, solid lines four-fold degeneracy and thick solid line eight-fold degeneracy of the state.

two electrons, the persistent current is doubled. The excited states can have varying degrees of degeneracy: If the quantum number pairs $P = [m_P, \tau_P]$ and $Q = [m_Q, \tau_Q]$ of the two electrons are identical then there is only one permitted configuration of the electron spins, the singlet state. However, if $P \neq Q$ then there are four degenerate possibilities: the singlet and three triplet states.

When the Coulomb interaction is included [Fig. 5.15 (b)], the picture changes drastically. To describe the two-particle states, we introduce the notation $M = m_1 + m_2$ for the total angular momentum and $T = \tau_1 + \tau_2$ for the total valley quantum number. The pair (M, T) denoting the two-particle state is differentiated from the single electron notation by the use of parentheses rather than brackets. The exchange interaction will split the degenerate singlet-triplet states when both of the electrons are in the same valley *i.e.* for $T = \pm 2$. In this case, the energy of the singlet does not contain any contribution from exchange and consequently has a rather higher energy than the corresponding triplet. This is exemplified by the $(M = 0, T = 2)$ state. The triplet part experiences exchange and this reduces its energy sufficiently for it to form the ground state for $\phi \approx 0.3$ with $\varepsilon \simeq 3.191$. At the same flux the singlet state has $\varepsilon \simeq 3.205$ and therefore has energy too high to be shown in Fig. 5.15 (b). On the other hand, the singlet and triplet parts of the $(-1, 0)$ degenerate state are not split by the exchange interaction.

For three non-interacting electrons in the ring (Fig. 5.16), the ground state is composed of spin and valley unpolarized states (*i.e.* $T = \pm 1$). When the interaction is added, the contribution from exchange is largest for $T = \pm 3$ states so the low energy spectrum becomes much more compact, just as in the $N = 2$ case. Qualitatively, the effect of the interaction is the same as previously, so that the changing nature of the ground state again demonstrates the complexity due to the absence of the valley degeneracy. However, because there are more possible combinations of states, the persistent current and absorption spectrum are correspondingly more complex in their structure. In particular it is not possible to have $T = 0$ so the exchange energy is always finite. However, its contribution is larger for $T = \pm 3$ states than for $T = \pm 1$ states. The interplay of the interaction and the total valley quantum number allow for an intricate manifestation of the breaking of valley degeneracy in this geometry. The change of the interacting ground state between singlet, triplet and degenerate singlet-triplet natures are manifested in the fractional nature of the AB oscillations in the persistent current, and in the steps and intensity changes in the absorption spectrum as the flux is varied.

A brief review of all the subsequent works on graphene QRs can be found in Schelter et al.[79]. Experimental realization of a graphene ring was first reported in 2008 [80]. Electron transport in that ring displayed clear AB oscillations in magnetoconductance.

[79]The Aharonov-Bohm effect in graphene rings, by J. Schelter, P. Recher, and B. Trauzettel, *Solid State Commun.* **152**, 1411 (2012).

[80]Observation of Aharonov-Bohm conductance oscillations in a graphene ring, by S. Russo, et al., *Phys. Rev. B* **77**, 085413 (2008).

5.6   Molecular adsorption on graphene

As we previously mentioned, pristine graphene does not have a band gap. This is in fact, a major obstacle for application of graphene in nanoelectronic devices, since the graphene transistor cannot be effectively turned off. There were various efforts to create an artificial band gap (band gap engineering) with limited success. For example, electrically biased bilayer graphene has a tunable band gap[81], but the gap is relatively small ~ 0.3 eV (mid infrared energy).

A promising route in this direction is the adsorption of various molecules on the graphene surface[82]. It has been suggested[83] that there is hardly any charge transfer between graphene and NH_3 when the H atoms point toward the graphene surface. But if the H atoms point away from the graphene surface then the charge transfer is $\sim 0.03e$, as in the latter case only the highest occupied molecular orbital (HOMO) of ammonia has a significant overlap with the graphene orbitals.

The band gap induced by the adsorption of NH_3, H_2O, CO, and HF have been studied by Berashevich and Chakraborty[84]. Those studies indicated that the resulting HOMO-LUMO (lowest unoccupied molecular orbital) gap can be up to ~ 1.5 eV -2.0 eV (Fig. 5.17). The adsorbed molecules on graphene are found to push the wave functions corresponding to up and down spin states to the opposite edges. This breaks the sublattice and molecular symmetries of graphene, thereby inducing a band gap. The efficiency of the wave function displacement depends on the type of molecules adsorbed (as shown in Fig. 5.17. The highest band gap due to water adsorption on graphene occurs when the dipole moment of the water molecules is directed toward the graphene. The gap is reduced to ~ 0.8 eV if the same dipole moment is directed away from

[81]Direct observation of a widely tunable bandgap in bilayer graphene, by Y. Zhang, et al., *Nature* **459**, 820 (2009).

[82]Molecular adsorption on graphene, by L. Kong, et al., *J. Phys.: Condens. Matter* **26**, 443001 (2014).

[83]Adsorption of H_2O, NH_3, CO, NO_2, and NO on graphene: A first-principles study, by O. Leenaerts, B. Partoens, and F.M. Peeters, *Phys. Rev. B* **77**, 125416 (2008).

[84]Tunable band gap and magnetic ordering by adsorption of molecules on graphene, by J. Berashevich and T. Chakraborty, *Phys. Rev. B* **80**, 033404 (2009).

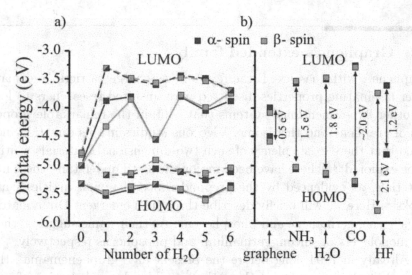

FIGURE 5.17
Effect of adsorption on the band gap in nanoscale graphene. (a) The HOMO and LUMO energies for graphene of size N = 3 (solid line) and N = 5 (dashed line) versus the number of adsorbed water molecules. (b) Influence on adsorption of different adsorbed molecules on the bandgap of the nanoscale graphene of size N = 4. Here N corresponds to the size of the graphene flake, i.e., the number of carbon rings along the zigzag edges (see Footnote 84).

the graphene. A band gap opening in graphene induced by the exposure to water molecules has indeed been observed in the experiments[85].

Large tunable band gaps were also reported for hydrogenated graphene[86]. Molecular adsorption on graphene studies are also important for graphene based sensors[87]. It could also be useful for the control of graphene geometry[88] and exploration of the magnetic properties of graphene[89].

---

[85]Tunable bandgap in graphene by the controlled adsorption of water molecules, by F. Yavari, et al., *Small* **6**, 2535 (2010).

[86]band gap tuning of hydrogenated graphene: H coverage and configuration dependence, by H. Gao, et al., *J. Phys. Chem.* **115**, 3236 (2011).

[87]Graphene-based gas sensors, by W. Yuan and G. Shi, *J. Mater. Chem.* **1**, 10078 (2013)

[88]Zipping and unzipping of nanoscale carbon structures, by J. Berashevich and T. Chakraborty, *Phys. Rev. B* **83**, 195442 (2011).

[89]Sustained ferromagnetism induced by H-vacancies in graphane, by J. Berashevich and T. Chakraborty, *Nanotechnology* **21**, 355201 (2010).

5.7   Graphene's extended family

Graphene , with massless Dirac fermions as charge carriers, and many
other fascinating properties discussed here, inspired researchers to look
for other two-dimensional systems that contain the remarkable properties of graphene and then some. Vigorous studies in this direction indicated that there exists plenty of such two-dimensional materials waiting
to be explored[90]. There have been major efforts to understand these new
materials, as evidenced by the preponderance of review articles[91] and
books[92]. Here we will briefly describe three such emergent Dirac materials, silicene, germanene and phosphorene. As their name suggests, these
are monolayers of silicon, germanium and phosphorus respectively.

Already in 1994, long before the present day 'graphenemania', theoretical studies were reported on the possible planar behavior of two-
dimensional layer of Si and Ge, where it was revealed that silicene
and germanene are actually buckled[93] instead of being fully planar as
graphene. This is due to larger radius of the Si/Ge atom compared to
the C atom, and as a result the two sublattices of the honeycomb lattices are slightly displaced in the direction vertical to the atomic plane.
These authors also found that silicene and germanene are semi-metals.
Later studies established that silicene and germanene share very similar
electronic properties with graphene, such as the Dirac cone near the K

---

[90]An atlas of two-dimensional materials, by P. Mir'o, M. Audiffred and T. Heine, *Chem. Soc. Rev.* **43**, 6537 (2014).

[91]Silicene and germanene: Silicon and germanium in the 'flatland', by A. Dimoulas, *Microelectronics Engineering* **131**, 68 (2015); Emergent elemental two-dimensional materials beyond graphene, by Y. Zhang, A. Rubio and G. Le Lay, *J. Phys. D: Appl. Phys.* **50**, 053004 (2017); Rise of silicene: A competitive 2D material, by J. Zhao, et al., *Prog. Mater. Sci.* **83**, 23 (2016); Silicene, a promising new 2D material, by H. Oughaddou, et al., *Prog. Surf. Sci.* **90**, 46 (2015); Germanene: the germanium ana-logue of graphene, by A. Acun, et al., *J. Phys.: Condens. Matter* **27**, 443002 (2015); Semiconducting black phosphorus: synthesis, transport properties and electronic ap-plications, by H. Liu, et al., *Chem. Soc. Rev.* **44**, 2732 (2015); The renaissance of black phosphorus, by X. Ling, et al., *PNAS* **112**, 4523 (2015).

[92]*Silicene: Prediction, Synthesis, Application*, edited by P. Vogt and G. Le Lay (Springer, 2018); *Silicene: Structure, Properties, and Applications*, edited by M.J.S. Spencer and T. Morishita (Springer, 2016).

[93]Theoretical possibility of stage corrugation in Si and Ge analogs of graphite, by K. Takeda and K. Shiraishi, *Phys. Rev. B* **50**, 14916 (1994).

and K' points of the Brillouin zone[94]. Both silicene and germanene suffer from the same malaise as graphene – a zero band gap. However, the buckling structure gives silicene and germanene large spin-orbit interactions which opens the band gaps at the Dirac points ($\Delta_{so} \approx 1.55 - 7.9$ meV) for silicene and ($\Delta_{so} \approx 24 - 93$ meV) for germanene. In the case of graphene, in contrast, the corresponding spin-orbit-induced gap is tiny (< 0.05 meV). Silicene and germanene do not exist in Nature in free standing form and unlike graphene, they cannot be exfoliated from the bulk sample. They are grown on metallic substrates[95]. The large spin-orbit interaction in these two systems is expected to enhance the fractional quantum Hall effect[96]. The Landau level energy spectrum and the quantum Hall states in buckled Dirac-like systems were found to differ significantly as compared to those in graphene[97].

The most promising member of the two-dimensional materials is phosphorene, a single layer of black phosphorus. Phosphorus was discovered in 1669 by a German alchemist Hennig Brandt[98]. In Nature, phosphorus is always found as phosphates, which is quite ubiquitous. It is in our bones and even in the DNA. Black phosphorus was discovered by Bridgman in 1914 in an attempt to 'force ordinary white phosphorus to change into red phosphorus by the application of high hydrostatic pressure, at a temperature below that at which the transformation runs with appreciable velocity at atmospheric pressure'[99]. Bridgman also noticed that unlike red and white phosphorus, black phosphorus (BP) is good electrical conductor and 'does not catch fire spontaneously', and 'can be ignited with difficulty with a match'. BP has a layered structure stacked together by van der Waals interactions. In each layer, there is a corrugated honeycomb structure where each P atom is covalently

[94]Two- and one-dimensional honeycomb structures of silicon and germanium, by S. Cahangirov, et al., *Phys. Rev. Lett.* **102**, 236804 (2009).

[95]Silicene: Compelling experimental evidence for graphenelike two-dimensional silicon, by P. Vogt, et al., *Phys. Rev. Lett.* **108**, 155501 (2012); Germanene: a novel two-dimensional germanium allotrope akin to graphene and silicene, by M.E. D'avila, et al., *New J. Phys.* **16**, 095002 (2014).

[96]Tuning of the Gap in a Laughlin-Bychkov-Rashba Incompressible Liquid, by M. Califano, T. Chakraborty, and P. Pietiläinen, *Phys. Rev. Lett.* **94**, 246801 (2005).

[97]Tunability of the fractional quantum Hall states in buckled Dirac materials, by V.M. Apalkov and T. Chakraborty, *Phys. Rev. B* **90**, 245108 (2014).

[98]A wonderful account of that discovery can be found in, The discovery of the elements. II. Elements known to the alchemists, by M.E. Weeks, *J. Chem. Edu.* **9**, 11 (1932).

[99]Two new modifications of Phosphorus, by P.W. Bridgman, *J. Am. Chem. Soc.* **36**, 1344 (1914).

bonded with three P atoms. Few-layer, and single layer BP can be obtained through mechanical exfoliation method akin to graphene[100]. The band gap in BP is 0.3 eV, and the gap increases to 1 - 2 eV, depending on the number of layers. The quantum Hall effect has been observed in the two-electron system in black phosphorus[101].

Single and bilayer phosphorene have been theoretically studied in the presence of perpendicular magnetic fields and the nature of the Landau levels were reported[102]. The fractional quantum Hall effect in phosphorene has been theoretically investigated[103]. The anisotropic band structure of phosphorene splits the collective excitations and additional modes appear in the spectrum. This system might provide a suitable playground to explore the geometrical picture of the FQHE where anisotropy plays a crucial role, as described in Sec. 2.8.

Undoubtedly, black phosphorus and phosphorene (among a few others) are the rising stars in the family of two-dimensional electron systems. Detailed accounts of the properties of many of those two-dimensional graphene-like materials can be found from various perspectives in recent publications[104].

5.8   Requiem for the (very expensive) dreams

When the simple 'scotch tape' method of producing graphene was presented to the public, there was electrifying response in the scientific community. There were high expectations that given all the

[100]Phosphorene: An unexplored 2D semiconductor with a high hole mobility, by H. Liu, et al., *ACS Nano* **8**, 4033 (2014).

[101]Quantum Hall effect in black phosphorus two-dimensional electron system, by L. Li, et al., *Nat. Nanotech.* **11**, 593 (2016).

[102]Landau levels and magnetotransport property of monolayer phosphorene, by X.Y. Zhou, et al., *Sci. Rep.* **5**, 12295 (2015); Landau levels of single-layer and bilayer phosphorene, by J.M. Pereira Jr. and M.I. Katsnelson, *Phys. Rev. B* **92**, 075437 (2015); Anomalous magneto-transport properties of bilayer phosphorene, by J.-Y. Wu, W.-P. Su, and G. Gumbs, *Sci. Rep.* **10**, 7674 (2020).

[103]Aspects of anisotropic fractional quantum Hall effect in phosphorene, by A. Ghazaryan and T. Chakraborty, *Phys. Rev. B* **92**, 165409 (2015).

[104]Two-dimensional graphene-like Xenes as potential topological materials, by A. Zhao and B. Wang, *APL Mater.* **8**, 030701 (2020); Elemental analogues of graphene: Silicene, germanene, stanene, and phosphorene, by S. Balendhrane et al., **11**, 640 (2014); The rare two-dimensional materials with Dirac cones, by J. Wang, et al., *National Sci. Rev.* **2**, 22 (2015).

spectacular properties promised (theoretically, at least) by graphene, it will usher in perhaps a new electronic revolution. Graphene was even touted by the media as the 'next wonder material'[105] that is poised to change the world. Huge funding was provided by the Governments to the researchers to deliver that promised industrial miracle. Interestingly, a clear winner in this race has been the publication industry. The subject of graphene has generated thousands upon thousands of publications by the researchers. New journals have appeared to cope with this avalanche of papers, review articles, numerous pedantic commentaries, news and views, etc. Publications are of course important for scientific advancement. As scholars have pointed out many decades ago, 'For most of us artisans of research, getting things into print becomes a symbolic equivalent to making a significant discovery. Nor could science advance without the great unending flow of papers reporting careful investigations, even if these are routine rather than distinctly original'[106]. Publications related to graphene have ostensibly validated that premise. However, concerns have been raised about the realistic market potential of graphene[107]. A graphene-based hair dye that adheres to the surface of hair, forming a coating that lasts more than 30 washes[108] might mean a world to a few, but it certainly does not qualify as a 'game changer' for graphene.

The breathtaking possibilities of graphene in nanoscale electronics have been clearly halted by the inability to mass produce this material in a monolayer form (having zero band gap did not help the case either). Production of strictly monolayers is essential since even having only a few layers of graphene would make it a graphite, whose practical applications are well established and will overshadow the extraordinary properties of graphene. What we need is to find a process to manufacture graphene in large quantities and in large enough sizes that could be integrated into electronic devices. Truly scalable production of graphene is still little more than just sheer grit. In addition to standard approaches to make graphene, some unconventional (but innovative) route has also

---

[105]https://www.theguardian.com/science/2013/nov/26/graphene-molecule-potential-wonder-material

[106]Priorities in scientific discovery: A chapter in the sociology of science, by R.M. Merton, *American Sociological Review* **22**, 635 (1957).

[107]Where does graphene go from here?, by D. Johnson, *IEEE Spectrum* **56**, 10 (2019); Graphene: the hype versus commercial reality, by T. Barkan, *Nat. Nanotechnol.* **14**, 904 (2019).

[108]Multifunctional graphene hair dye, by C. Luo, et al., *Chem* **4**, 784 (2018).

been explored. There were reports of making graphene from dog feces[109] and bird droppings[110]. There is even a war raged on production of *fake graphene*[111] for commercial purposes. Clearly, the wonder material has failed as yet, to satisfy the high expectations levied on it.

Historically, the label 'wonder material' and elevated hope and the hype about a particular substance have sometimes led to inevitable disappointments. In this context, the famous story of the ups and downs in public's appreciation of aluminum (or aluminium) comes to mind[112]. Although in use since ancient times, only in 1533 a Swiss-German physician and alchemist named Philippus Aureolus Theophrastus Bombastus von Hohenheim studied and wrote about the 'alum earth' containing the oxide of an unknown metal, subsequently called alumina. Then, in 1754, a German chemist Andreas Sigismund Marggraf synthesized alumina from clay. Although the English chemist Sir Humphry Davy could not isolate the elusive metal, he named it aluminum (and later aluminium) in the early nineteenth century. In fact, it was the turn of the German chemist Friedrich Wöhler to prepare the first pure sample of aluminum in 1827, albeit in the form of a few grains.

Since the metal was so scarce and difficult to extract, naturally back then it was more expensive than gold. In 1855, the 'silver from clay' caused a huge sensation, mostly among the royalties, when it was displayed at the world fair in Paris. It was such a sought-after metal that, as the legend has it, the then Emperor Napoleon III once gave a banquet where the members of the royal family and many dignitaries were all given aluminum forks and spoons to dine with, while others had to be content having to use the ordinary gold and silver cutlery. However, over time (toward the end of nineteenth century) large scale extraction of aluminum was made possible, and most of its wonderful properties for practical use were established. Since it was readily available in large quantities, quite naturally, the price of aluminum fell considerably, and so dropped the allure of this wonder material. Today, between soft-drink cans to food wrappings to aircraft fuselage, aluminum plays important roles in countless objects we use in our daily life. But one thing is

---

[109]Growth of graphene from food, insects, and waste, by G. Ruan, et al., *ACS Nano* **5**, 7601 (2011).

[110]Will any crap we put into graphene increase its electrocatalytic effect?, by L. Wang, et al., *ACS Nano* **14**, 21 (2020).

[111]The war on fake graphene, by P. Boggild, *Nature* **562**, 502 (2018).

[112]*This story was wonderfully told in 'Tales About Metals',* by S.I. Venetsky (Mir Publishers, Moscow, 1981).

certain, a royal feast with the cutleries made of that wonder material will definitely be out of question! Coming back to graphene in this century, one can only hope that the billions of dollars spent and thousands of researchers working furiously to mass produce today's wonder material will finally result in spectacular devices that will change our lives forever.

6

Some remarkable episodes in the nanoscale

In this chapter, we describe a few phenomena that have remained intriguing theoretical fascinations for many years but are only recently realized (albeit partially) in nanoscale systems. Among those are the fractal butterflies, Maxwell's intelligent demon, and the electronic properties of DNA. Although they look like totally disconnected topics they are all closely related to the subjects described in previous chapters.

6.1 Fractal butterflies

Among all the insects, the butterflies are perhaps the most revered in human culture. Since ancient times, their fleeting flight and metamorphosis, not to mention their bright colored and patterned wings, have fascinated people and these winged beautiful creatures have been celebrated in paintings, literature, and photography. Aristotle named the butterfly *psyche*, the Greek word for soul, and in many societies, the butterfly has indeed been a symbol for the human soul. Now it is the turn of the scientists, other than the lepidopterists to explore the beautiful butterflies. The fractal butterflies, to be described below, have fascinated physicists and mathematicians for a few decades in order to understand the intricate patterns the mathematically created butterflies make, and eventually provide a rare glimpse of the structure in the experiments.

The dynamics of an electron in a periodic potential and subjected to a perpendicular magnetic field has remained a fascinating topic for physicists (and also for mathematicians to some extent) for more than half a century. Within the nearest neighbor tight-binding description of the periodic potential, the energy spectrum of an electron is described by

DOI: 10.1201/9781003090908-6

the so-called Harper's equation[1] (see Box 6.1). The energy spectra determined by the Harper equation is a periodic function of the dimensionless parameter $\tilde{\alpha}$ (the number of flux quanta through a unit cell) with period 1. Hence it is sufficient to consider only the values of $\tilde{\alpha}$ within the range $0 < \tilde{\alpha} < 1$. The remarkable property of the Harper equation is that although the corresponding Hamiltonian is an analytical function of $\tilde{\alpha}$, the energy spectrum is very sensitive to the choice of $\tilde{\alpha}$. At rational values of $\tilde{\alpha} = p/q$ (p and q being integers, and p/q is an irreducible fraction) the energy spectrum has q bands separated by $q-1$ gaps, where each band is p-fold degenerate. On the other hand, for irrational $\tilde{\alpha}$ the spectrum consists of infinitely many isolated points. In his famous 1976 paper[2] Hofstadter showed that the energy spectrum has a *fractal* structure[3], i.e., a structure that is same as itself at different scales. In other words, a detailed pattern that repeats itself ad infinitum, the 'romanesco' in the field of mathematics, so to speak. The energy spectrum in the present context is known in the literature as Hofstadter's butterfly (because of the pattern resembling the butterflies). This is the first example of the fractal pattern realized in the energy spectra of a quantum system. The spectrum is shown in Fig. 6.1[4].

Box 6.1 The Harper equation:

The dynamics of a 2D electron in a periodic potential is described by the Hamiltonian: $\mathcal{H} = \mathcal{H}_0\left(p_x, p_y\right) + V(x, y)$, where the first term on the right-hand side is the kinetic energy and the second term is the periodic potential. In the presence of a magnetic field and choosing the vector potential as $\mathbf{A} = (0, Bx)$, i.e., in the Landau gauge, the corresponding Hamiltonian is, $\mathcal{H} = \mathcal{H}_0\left(p_x, p_y - exB\right) + V(x, y)$. We consider the two limiting cases of weak and strong magnetic fields. *Weak magnetic field:* First the periodic potential forms the Bloch

---

[1]Single band motion of conduction electrons in a uniform magnetic field, by P.G. Harper, *Proc. Phys. Soc.* A **68**, 874 (1955); The tight-binding and the nearly-free-electron approach to lattice electrons in external magnetic fields, by D. Langbein, *Phys. Rev.* **180**, 633 (1969).

[2]Energy levels and wave functions of Bloch electrons in rational and irrational magnetic fields, by D. Hofstadter, *Phys. Rev. B* **14**, 2239 (1976).

[3]The fractal geometry of nature, by B.B. Mandelbrot, (W.H. Freeman and Company, New York, 1982).

[4]Fractal butterflies of Dirac fermions in monolayer and bilayer graphene, by T. Chakraborty and V.M. Apalkov, *IET Circuits Devices Syst.* **9**, 19 (2015).

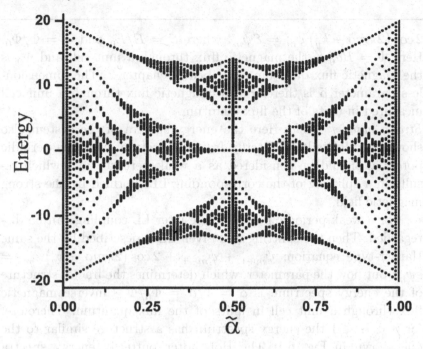

FIGURE 6.1
Energy spectrum (Hofstadter butterfly) of an electron in a periodic potential and a perpendicular magnetic field, as a function of the magnetic flux per unit cell in units of the flux quantum.

bands and the magnetic field splits each Bloch band into mini-bands of the LL type. For a weak magnetic field, the coupling of different bands can be disregarded. The corresponding Schrödinger equation, which determines the energy spectrum of the system, has a simple form in the tight-binding approximation of the periodic potential for which the energy dispersion within a single band is[5] $E(p_x, p_y) = 2E_0 \left(\cos(p_x a_0/\hbar) + \cos(p_y a_0/\hbar) \right)$, where a simple square lattice strcuture with lattice constant a_0 was assumed. Here E_0 is the nearest-neighbor hopping amplitude.

In an external magnetic field, the wave function which is defined at the lattice points (ma_0, na_0), has the form, $\Psi(ma, na) = e^{ik_y n}\psi_m$. The corresponding Schrödinger equation reduces to a 1D equation – the so-called Harper equation, $\psi_{m+1} + \psi_{m-1} +$

$2\cos\left(2\pi m\tilde{\alpha} - k_y\right)\psi_m = \mathcal{E}\psi_m$, where $\mathcal{E} = E/E_0$ and $\tilde{\alpha} = \Phi/\Phi_0$. Here $\Phi = Ba_0^2$ is the magnetic flux through a unit cell and Φ_0 is the magnetic flux quantum described in chapter 2. The dimensionless parameter $\tilde{\alpha}$ is therefore the magnetic flux through a unit cell measured in units of the flux quantum.

Strong magnetic field: Here the energy spectra of the system also shows the Hofstadter butterfly fractal structure. Now the periodic potential should be considered as a weak perturbation, which results in a splitting of the corresponding LLs, formed by the strong magnetic field.

For a weak periodic potential the inter LL coupling can be disregarded. Then the splitting of a given LL is described by the same Harper-type equation: $\psi_{m+1} + \psi_{m-1} + 2\cos\left(2\pi m\alpha - k_y\right)\psi_m = \varepsilon\psi_m$, but now the parameter, which determines the fractal structure of the energy spectrum, is $\alpha = 1/\tilde{\alpha} = \Phi_0/\Phi$ – inverse magnetic flux through a unit cell in units of the flux quantum. Therefore, for $0 < \alpha < 1$ the energy spectrum has a structure similar to the one shown in Fig. 6.1. The Hofstadter butterfly energy spectra is realized either as a splitting of the Bloch band by a weak magnetic field or as a splitting of a LL by the weak periodic potential.

Experimental efforts to detect the butterfly pattern in the energy spectrum of an electron in a periodic lattice and a perpendicular magnetic field are quite a daunting task. Finding the self-similar pattern is inherently limited due to the finite resolution of the instruments and the accessibility of the required magnetic fields. For example, in a typical crystalline lattice with lattice spacing of ~ 0.1 nm, a magnetic field of about 10^5 Tesla would be required to realize the butterfly pattern. This is far beyond the accessible magnetic fields. Clearly, artificial lattices with larger periods (as opposed to those in natural crystals) will be required to keep the magnetic field in a reasonable range of values. Measurement of quantized Hall conductance in artificial semiconductor lattice

[5]Bloch electrons in a magnetic field: Hofstadter's butterfly, by U. Rössler and M. Shurke, in *Advances in Solid State Physics*, edited by B. Kramer (Springer Heidelberg 2000), vol. 40, p. 35.

structures[6] indicated, albeit indirectly, the complex pattern of gaps that were expected in the butterfly spectrum. Butterfly patterns are also expected to appear in propagation of microwaves through a waveguide with a periodic array of scatteres[7].

In Box 6.1, we have considered the cases of weak and strong magnetic fields. For intermediate values of the magnetic field, the mixing of the LL by the periodic potential or the mixing of Bloch bands by the magnetic field becomes strong. This will modify the butterfly structure and add some system-dependent features. In what follows, we consider the limits of high and intermediate magnetic fields for conventional semiconductor systems and for monolayer and bilayer graphene.

6.1.1 Semiconductor systems: Strong field limit

For strong and intermediate magnetic fields, the periodic potential is considered as a perturbation, which can modify and mix the states of the zero-order Hamiltonian, consisting of only the kinetic energy $\mathcal{H}_0 \left(p_x - eA_x, p_y - eA_y\right)$. For conventional semiconductor systems, the zero-order Hamiltonian is described by the parabolic dispersion relation, $p^2/2m$. The transverse magnetic field results in Landau quantization where the LLs are characterized by the LL index, $n = 0, 1, 2, \ldots$ with energies $E_n = \left(n + \frac{1}{2}\right)\hbar\omega_c$, where $\omega_c = eB/m$ is the cyclotron frequency (see Chapter two). The corresponding Landau wave functions are: $\varphi_{n,k}(x,y) = \frac{e^{iky}}{\sqrt{L}} \frac{e^{-(x-x_k)^2/2\ell_0^2}}{\sqrt{\pi^{1/2}\ell_0 2^n n!}} \mathrm{H}_n\left(x - x_k\right)$, where L is the length of a sample in the y direction, k is the y component of the electron wave vector, $\ell_0 = \sqrt{\hbar/eB}$ is the magnetic length, $x_k = k\ell_0^2$, and $\mathrm{H}_n(x)$ are the Hermite polynomials. We consider the system in a periodic external potential of the form, $V(x,y) = V_0\left[\cos\left(q_x x\right) + \cos\left(q_y y\right)\right]$, where V_0 is the amplitude of the periodic potential, $q_x = q_y = q_0 = 2\pi/a_0$, and a_0 is a period of the external potential $V(x,y)$. The expressions for the matrix elements of the periodic potential $V(x,y)$ in the basis $\phi_{n,k}(x,y)$ are given in Box 6.2.

---

[6]Detection of Landau band coupling induced rearrangement of the Hofstadter butterfly, by M.C. Geisler, et al., *Physica E* **25**, 227 (2004); Evidence of Hofstadter's fractal energy spectrum in the quantized Hall conductance, by C. Albrecht, et al., *Physica E* **20**, 143 (2003); Landau subbands generated by a lateral electrostatic superlattice – chasing the Hofstadter butterfly, by T. Schlösser, et al., *Semicond. Sci. Technol.* **11**, 1582 (1996).

[7]Microwave realization of the Hofstadter butterfly, by U. Kuhl and H.-J. Stöckmann, *Phys. Rev. Lett.* **80**, 3232 (1998).

In general, the expressions for the matrix elements (6.2) and (6.1) can be used to find the energy spectra of any finite number of LLs, taking into account the coupling of different LLs introduced by the periodic potential. For a given value of k within the interval $0 < k < q_0$, a finite set of basis wave functions $\phi_{n,k}$, $\phi_{n,k+q_0}$, $\phi_{n,k+2q_0}$,..., $\phi_{n,k+N_x q_0}$ is considered. Here $n = 0, \ldots, N_L$, N_L is the number of LLs, and N_x determines the size of the system in the x direction: $L_x = N_x q_0 \ell_0^2$. The matrix elements (6.1) and (6.2) determine the coupling of the states within this truncated basis and finally determine the corresponding Hamiltonian matrix. The diagonalization of the matrix provides the energy spectrum for a given value of k. The spectra are calculated for a finite number N_y of k points, where N_y determines the size of the system in the y direction: $L_y = 2\pi N_y / q_0$.

Following this procedure the energy spectra of the conventional system with parabolic dispersion relation were evaluated for $N_L = 2$ LLs. The results are shown in Fig. 6.2 for the period of the potential $a_0 = 20$ nm. The results clearly show that although for the potential amplitude $V_0 = 10$ meV the mixing of LLs is relatively weak, for a higher amplitude $V_0 = 20$ meV the mixing becomes strong especially for α close to 1, i.e., in weak magnetic fields. The butterfly structure is no longer described by the simple Harper equation. A detailed analysis of the Hofstadter butterfly spectrum for strong and intermediate periodic potential strength indicates that[8] the magnetic field splits the Bloch bands and introduces coupling of the states of different Bloch bands.

Box 6.2 The matrix elements (semiconductor)

The matrix elements of the periodic potential $V(x,y)$ in the basis $\phi_{n,k}(x,y)$ are

$$\langle \phi_{n',k'} | \cos(q_0 y) | \phi_{n,k} \rangle =$$
$$\frac{i^{|n'-n|-(n'-n)}}{2} \left\{ \delta_{k',k+q_0} + (-1)^{n-n'} \delta_{k',k-q_0} \right\} M_{n',n} \quad (6.1)$$

---

[8]Two-dimensional Bloch electrons in perpendicular magnetic fields: an exact calculation of the Hofstadter butterfly spectrum, by S. Janecek, M. Aichinger, and E.R. Hernandez, *Phys. Rev. B* **87**, 235429 (2013).

and

$$\langle \phi_{n',k'} | \cos(q_0 x) | \phi_{n,k} \rangle =$$

$$i^{|n'-n|} \frac{\delta_{k',k}}{2} \left[e^{iq_0 k \ell_0^2} + (-1)^{n-n'} e^{-iq_0 k \ell_0^2} \right] M_{n',n}. \quad (6.2)$$

Here

$$M_{n',n} = \left(\frac{m!}{M!} \right)^{1/2} e^{-Q/2} Q^{|n'-n|/2} L_m^{|n'-n|}(Q), \quad (6.3)$$

$Q = q_0^2 \ell_0^2 / 2$, $m = \min(n', n)$, $M = \max(n', n)$, and L_m^n is the generalized Laguerre polynomial.

These matrix elements are used in the construction of the Hamiltonian matrix for evaluation of the energy spectra as a function of α.

The matrix elements (6.1) and (6.2) (see Box 6.2) determine the mixture of the LL states introduced by the periodic potential. While the component of the potential periodic in the x direction [Eq. (6.2)] couples only the states with the same value of the wave vector k, the component periodic in the y direction couples the states with the wave vectors separated by q_0. Within a single LL the potential periodic in the x direction modifies the energy of each Landau state. As a result the energy of the Landau state within a given LL becomes a periodic function of $q_0 k \ell_0^2$. Additional coupling of the states separated by q_0, which is determined by Eq. (6.1), results in the formation of the band structure when $q_0^2 l_0^2$ becomes a rational fraction of 2π, which is exactly the condition that the parameter α is rational. It follows from Eqs. (6.1)-(6.2) that for a given LL with index n the effective amplitude of the periodic potential acquires an additional factor and becomes $\propto V_0 M_{n,n} \propto L_n(q_0^2 \ell_0^2 / 2) = L_n(\pi \alpha)$. Here L_n is the Laguerre polynomial. These renormalized amplitudes determine the width of the corresponding bands. At values of α where $L_n(\pi \alpha) = 0$, all bands have zero width which correspond to the flatband condition[9].

6.1.2 Butterflies in monolayer graphene

Graphene with its unusual electronic properties, described in the previous Chapter, has turned out to be the ideal system in the quest of fractal

[9]Detection of a Landau band-coupling-induced rearrangement of the Hofstadter butterfly, by M.C. Geisler, et al., *Phys. Rev. Lett.* **92**, 256801 (2004).

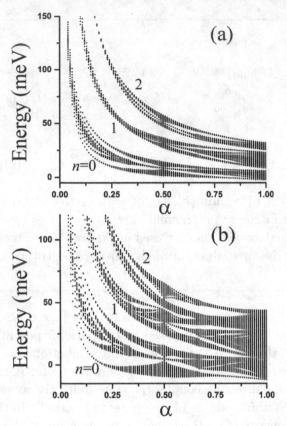

FIGURE 6.2
Single-electron energy spectra of conventional semiconductor systems
with parabolic dispersion relation. The period of the potential is $a_0 = 20$
nm and its amplitude is (a) $V_0 = 10$ meV and (b) $V_0 = 20$ meV. The
energy spectra are shown as a function of the parameter $\alpha = \Phi_0/\Phi$. The
numbers indicate the LL index n.

butterflies. The Dirac fermions in monolayer and bilayer graphene are
the most promising objects thus far, where the signature of the recursive
pattern of the Hofstadter butterfly has been unambiguously reported[10].

---

[10]Hofstadter's butterfly and the fractal quantum Hall effect in Moiré superlattices,
by C.R. Dean, et al., *Nature* **497**, 598 (2013); Massive Dirac fermions and Hofstadter
butterfly in a van der Waals heterostructure, by B. Hunt, et al., *Science* **340**, 1427
(2013); Cloning of Dirac fermions in graphene superlattices, by L.A. Ponomarenko,
et al., *Nature* **497**, 594 (2013).

Here the periodic lattice with a period of ~ 10 nm was created by the Moiré pattern that appears when graphene is placed on a plane of hexagonal boron nitride (h-BN) with a twist[11]. Being ultraflat and free of charged impurities, h-BN has been the best substrate for graphene having high-mobility charged fermions. Some theoretical studies have been reported earlier in the literature on the butterfly pattern in monolayer[12] and bilayer graphene[13].

6.1.3 Square lattice periodic structure

As discussed in the previous Chapter, the LLs in graphene have two-fold valley degeneracy corresponding to two valleys K and K'. The degeneracy cannot be lifted by the periodic potential with typical long periods, $a_0 > 10$ nm. In this case the Hofstadter butterfly pattern realized in graphene have two-fold valley degeneracy and it is enough to consider only the states of one valley, e.g., valley K. The corresponding Hamiltonian $\mathcal{H}_0$ is written in the matrix form (see Chapter 5)

$$\mathcal{H}_0 = v_F \begin{pmatrix} 0 & \pi_x - i\pi_y \\ \pi_x + i\pi_y & 0 \end{pmatrix}, \tag{6.4}$$

where $\vec{\pi} = \vec{p} + e\vec{A}/c$, $\vec{p}$ is the electron momentum and $v_F \approx 10^6$ m/s is the Fermi velocity.

The LLs in graphene, which are determined by the Hamiltonian (6.4), are specified by the Landau index $n = 0, \pm 1, \pm 2, \ldots$, where the positive and negative values correspond to the conduction and valence band levels, respectively. The energy of the LL with index n is

$$E_n^{(gr)} = s_n \hbar \omega_{gr,B} \sqrt{|n|}, \tag{6.5}$$

where $\omega_{gr,B} = v_F/\ell_0$ is the cyclotron frequency in graphene; $s_n = 1$ for $n > 1$, $s_n = 0$ for $n = 0$, and $s_n = -1$ for $n < 1$.

The eigenfunctions of the Hamiltonian (6.4), corresponding to the LL with index n, are given by

$$\Psi_{n,k} = C_n \begin{pmatrix} s_n i^{|n|-1} \phi_{|n|-1,k} \\ i^{|n|} \phi_{|n|,k} \end{pmatrix}, \tag{6.6}$$

---

[11]Boron nitride substrates for high-quality graphene electronics, by C.R. Dean, et al., *Nat. Nanotech.* **5**, 722 (2010).

[12]Self-similar occurrence of mass Dirac particles in graphene under a magnetic field, by J.-W. Rhim and K. Park, *Phys. Rev. B* **86**, 235411 (2012).

[13]Hofstadter butterflies of bilayer graphene, by N. Nemec and G. Cuniberti, *Phys. Rev B* **75**, 201404(R) (2007).

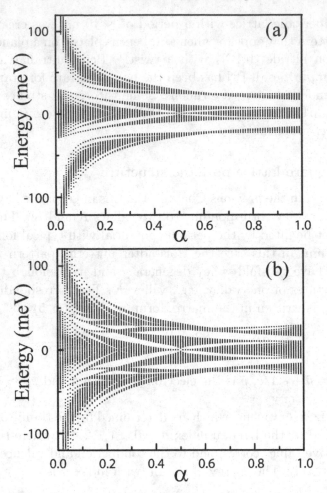

FIGURE 6.3

Single-electron energy spectra of graphene monolayer in a periodic potential and an external magnetic field. The period of the potential is $a_0 = 20$ nm and its amplitude is (a) $V_0 = 50$ meV and (b) $V_0 = 100$ meV. The energy spectra are shown as a function of $\alpha = \Phi_0/\Phi$.

where $C_n = 1$ for $n = 0$ and $C_n = 1/\sqrt{2}$ for $n \neq 0$. Here $\phi_{n,k}$ is the Landau wave function introduced above for an electron with parabolic dispersion relation. The graphene monolayer is then placed in a weak periodic potential $V(x, y)$, introduced above. This potential causes coupling of LLs in graphene.

Box 6.3 The matrix elements (graphene)

The corresponding matrix elements of the periodic potential in graphene are

$$\langle n'k'| \cos(q_0 y) |nk\rangle =$$
$$\frac{i^{n-n'}}{2} C_n C_{n'} \left\{ \delta_{k',k+q_0} + (-1)^{n-n'} \delta_{k',k-q_0} \right\} \times$$
$$\left[s_n s_{n'} M_{|n'|-1,|n|-1} + M_{|n'|,|n|} \right] \tag{6.7}$$

and

$$\langle n'k'| \cos(q_0 x) |nk\rangle = \frac{\delta_{k',k}}{2} C_n C_{n'} e^{-iq_0 k \ell_0^2} \tag{6.8}$$
$$\times \left[1 + (-1)^{n-n'} \right] \left[s_n s_{n'} M_{|n'|-1,|n|-1} + M_{|n'|,|n|} \right].$$

For a given LL with index n, the periodic potential is determined by the effective value

$$V_0 \left[s_n^2 M_{|n|-1,|n|-1} + M_{|n|,|n|} \right]$$
$$\propto V_0 \left[s_n^2 L_{|n|-1}(\pi\alpha) + L_{|n|}(\pi\alpha) \right]. \tag{6.9}$$

The flatbands in graphene are therefore realized at points where $s_n^2 L_{|n|-1}(\pi\alpha) + L_{|n|}(\pi\alpha)$ is 0. For $n = 0$, i.e., $s_0 = 0$, this is exactly the same condition as in conventional system, but for other LLs the condition of flatbands becomes $L_{|n|-1}(\pi\alpha) + L_{|n|}(\pi\alpha) = 0$. Here, $L_m(x)$ are the Laguerre polynomials.

In Fig. 6.3 the Hofstadter butterfly energy spectra is shown for a graphene monolayer, taking into account three LLs with $n = -1, 0$, and 1. The main difference between the conventional systems and graphene is the broadening of the energy structure within a single LL. For conventional systems [Fig. 6.2], the width of the energy spectra for the $n = 1$ LL is small for small values of α and large for large α. In graphene the behavior is different: the broadening of the $n = 1$ LL is large for small

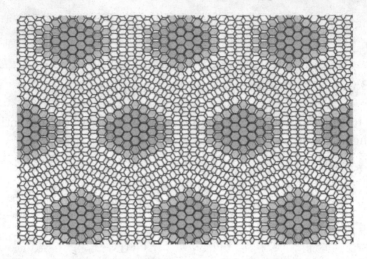

FIGURE 6.4

Moiré pattern in two hexagonal lattices with rotational misalignment: Image of the pattern for two graphene layers stacked with a rotation angle of 5°. The shaded parts denote the areas where the two lattices are almost superposed.

values of α and small for intermediate and large values of α. Another specific feature of the energy spectra of graphene is that the mixing of the LLs, introduced by the periodic potential, is visible for much larger values of the amplitude of the potential, $V_0 \approx 100$ meV compared to $V_0 \approx 20$ meV in conventional systems [Fig. 6.2 (b)].

6.1.4 Moiré structure

The moiré pattern is a well known phenomenon which appears when repetitive structures such as screens or grids are superposed or viewed against each other with a little rotation[14]. It appears as a new pattern of alternating dark and bright areas which is clearly observed at the superposition, although they are not present in any of the original structures. In graphene we have the unique possibility to generate such a pattern due to the intrinsic structure of graphene-based systems. In the case of graphene, the Moiré pattern is realized in a graphene monolayer placed on a hexagonal boron nitride (h-BN) substrate with *rotational*

---

[14] *The Theory of the Moiré Phenomenon*, by I. Amirdror (Springer, London 2009).

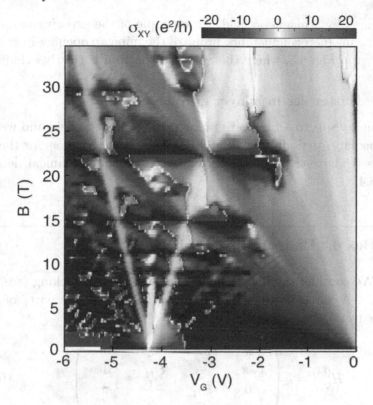

FIGURE 6.5
Experimental results for the Hall conductance probe of minigap opening within a Landau level in graphene, depicting the self-similar pattern (See Footnote 4).

misalignment between the graphene monolayer and the h-BN [15]. An example of the Moiré pattern in two hexagonal lattices (layers)[16] is shown in Fig. 6.4. That pattern introduces a large-scale periodicity in the Hamiltonian of the systems, which in a magnetic field could facilitate observation of the Hofstadter butterfly spectra.

Formation of the fractal butterfly pattern in graphene on the h-BN substrate was indeed observed experimentally (see Footnote 10) by a few groups. This butterfly pattern was realized as splitting of the Moiré

---

[15]See Footnote 10.

[16]Interlayer repulsion and decoupling effects in stacked turbostratic graphene flakes, by J. Berashevich and T. Chakraborty, *Phys. Rev. B* **84**, 033403 (2011).

minibands by a magnetic field. An example of the experimental results from the magnetoconductance probe of the minigap opening in graphene is shown in Fig. 6.5, where the repetitive pattern is clearly evident.

6.1.5 Butterflies in bilayer graphene

The energy spectra of bilayer graphene in a magnetic field and weak (or intermediate) periodic potential follows the same route as for the cases discussed above. For completeness, we discuss the technical details in Box. 6.4.

Box 6.4 Basics of (biased) bilayer graphene:

We consider the bilayer graphene with Bernal stacking (see Sec. 5.4.2). A single-particle Hamiltonian (kinetic energy part) of this system in a magnetic field is[17]

$$\mathcal{H}_\xi^{(bi)} = \xi \begin{pmatrix} \frac{U}{2} & v_F \pi_- & 0 & 0 \\ v_F \pi_+ & \frac{U}{2} & \xi \gamma_1 & 0 \\ 0 & \xi \gamma_1 & -\frac{U}{2} & v_F \pi_- \\ 0 & 0 & v_F \pi_+ & -\frac{U}{2} \end{pmatrix}, \qquad (6.10)$$

where $\xi = \pm 1$ corresponds to two valley (K and K'), U is the inter-layer bias voltage which can be varied for a given system, and $\gamma_1 \approx 0.4$ eV is the inter-layer coupling. The eigenfunctions of the Hamiltonian (6.10) can be expressed in term of the Landau functions $\phi_{n,k}$ (see Sec. 2.1.1)

$$\Psi_{n,k}^{(bi)} = \begin{pmatrix} \xi C_1 \phi_{|n|-1,k} \\ i C_2 \phi_{|n|,k} \\ i C_3 \phi_{|n|k} \\ \xi C_4 \phi_{|n|+1,k} \end{pmatrix}, \qquad (6.11)$$

where the coefficients, C_1, C_2, C_3, and C_4, can be found from the following system of equations

$$\varepsilon C_1 = \xi u C_1 - \sqrt{n} C_2 \qquad (6.12)$$
$$\varepsilon C_2 = \xi u C_2 - \sqrt{n} C_1 + \tilde{\gamma}_1 C_3 \qquad (6.13)$$
$$\varepsilon C_3 = -\xi u C_3 + \sqrt{n+1} C_4 + \tilde{\gamma}_1 C_2 \qquad (6.14)$$
$$\varepsilon C_4 = -\xi u C_4 + \sqrt{n+1} C_3. \qquad (6.15)$$

Here all energies are expressed in units of $\epsilon_B = \hbar v_F / \ell_0$, ε is the energy of the LL, $u = U/(2\epsilon_B)$, and $\tilde{\gamma}_1 = \gamma_1/\epsilon_B$. The energy spectra of the LLs can be found from[18]

$$\left[(\varepsilon + \xi u)^2 - 2n \right] \left[(\varepsilon - \xi u)^2 - 2(n+1) \right] = \tilde{\gamma}_1^2 \left[\varepsilon^2 - u^2 \right]. \qquad (6.16)$$

For each value of $n \geq 0$ there are four solutions of the eigenvalue equation (6.16), corresponding to four Landau levels in a bilayer graphene for a given valley, $\xi = \pm 1$. For zero bias voltage, $U = 0$ these four Landau levels are

$$\epsilon = \pm \sqrt{2n + 1 + \frac{\tilde{\gamma}_1^2}{2} \pm \frac{1}{2} \sqrt{(2 + \tilde{\gamma}_1^2)^2 + 8n\tilde{\gamma}_1^2}}. \qquad (6.17)$$

In this case each Landau level has two-fold valley degeneracy which is lifted at finite bias voltage U.

For $n = 0$ there are two special LLs of bilayer graphene. One LL has the energy $\varepsilon = -\xi u$ and the wave function of this LL consists of $\phi_{0,k}$ functions only

$$\Psi_{0_1,k}^{(bi)} = \begin{pmatrix} \phi_{0,k} \\ 0 \\ 0 \\ 0 \end{pmatrix}. \qquad (6.18)$$

This LL of bilayer graphene has exactly the same properties as for the 0-th conventional, non-relativistic Landau level. For zero bias voltage U, this level has zero energy.

For small values of U there is another solution of Eq. (6.16) with $n = 0$, which has almost zero energy, $\varepsilon \approx 0$. The corresponding LL has the wavefunction

$$\Psi_{0_2,k}^{(bi)} = \frac{1}{\sqrt{\gamma_1^2 + 2\epsilon_B^2}} \begin{pmatrix} \gamma_1 \phi_{1,k} \\ 0 \\ \sqrt{2}\epsilon_B \phi_{0,k} \\ 0 \end{pmatrix}. \qquad (6.19)$$

The wave function of this LL is the mixture of the $n = 0$ and $n = 1$ conventional ('nonrelativistic' in graphene parlance) Landau functions $\phi_{0,k}$ and $\phi_{1,k}$. This mixing depends on the magnitude of the magnetic field. In a small magnetic field $\epsilon_B \ll \gamma_1$, the wavefunction is $(\psi_{1,m}, 0, 0, 0)^T$ and the LL is identical to the $n = 1$ nonrelativistic LL. In a large magnetic field $\epsilon_B \gg \gamma_1$, the LL wavefunction is $(0, 0, \psi_{0,m}, 0)^T$ and the bilayer LL has the same properties as the $n = 0$ non-relativistic LL.

Following the same procedure as for the conventional systems and the graphene monolayer, we can find the matrix elements of the periodic potential in the basis of LL wave function of bilayer graphene (Box 6.5).

Box 6.5 The matrix elements (Bilayer graphene):

The matrix elements of the periodic potential in this case are:

$$\langle n'k' | \cos(q_0 y) | nk \rangle =$$

$$\frac{i^{n-n'}}{2} C_n C_{n'} \left\{ \delta_{k',k+q_0} + (-1)^{n-n'} \delta_{k',k-q_0} \right\}$$

$$\times \Big[C_{n,1} C_{n',1} M_{|n'|-1,|n|-1} + C_{n,4} C_{n',4} M_{|n'|+1,|n|+1}$$

$$+ \left(C_{n,2} C_{n',2} + C_{n,3} C_{n',3} \right) M_{|n'|,|n|} \Big] \tag{6.20}$$

and

$$\langle n'k' | \cos(q_0 x) | nk \rangle = \frac{\delta_{k',k}}{2} C_n C_{n'} \, e^{-i q_0 k \ell_0^2} \left[1 + (-1)^{n-n'} \right]$$

$$\times \Big[C_{n,1} C_{n',1} M_{|n'|-1,|n|-1} + C_{n,4} C_{n',4} M_{|n'|+1,|n|+1}$$

$$+ \left(C_{n,2} C_{n',2} + C_{n,3} C_{n',3} \right) M_{|n'|,|n|} \Big] . \tag{6.21}$$

These equations are all that is required to evaluate the energy spectra.

---

[17]Landau-level degeneracy and quantum Hall effect in a graphene bilayer, by E. McCann and V. Falko, *Phys. Rev. Lett.* **96**, 086805 (2006); Asymmetry gap in the electronic band structure of bilayer graphene, by E. McCann, *Phys. Rev. B* **74**, 161403 (2006).

[18]Landau levels and oscillator strength in a biased bilayer of graphene, by M.J. Pereira, F.M. Peeters, and P. Vasilopoulos, *Phys. Rev. B* **76**, 115419 (2007).

With the known matrix elements of the periodic potential, we can find the energy spectra of bilayer graphene in a magnetic field and weak (or intermediate) periodic potential, taking into account many LLs. The results are shown in Fig. 6.6. For zero bias voltage [Fig. 6.6(a)], similar to monolayer graphene, the inter-Landau level coupling becomes important only for large amplitudes of the periodic potential, $V_0 > 100$ meV. This is true except for two degenerate LLs of type (6.18) and (6.19), for which the inter-level coupling becomes strong even for small amplitudes V_0 due to the degeneracy of the levels. In this case, the structure of the energy spectrum near zero energy becomes complicated due to the mixture of two degenerate butterfly structures. These two butterfly structures are not identical due to different types of wave functions of the two LLs and correspondingly different effective periodic potentials. For one LL the effective periodic potential is $V_0 L_0(\pi\alpha)$, while for the other LL, the wave function of which is given by Eq. (6.19), the effective strength of the potential is

$$\frac{V_0}{\gamma_1^2 + 2\epsilon_B^2} \left(\gamma_1^2 L_1(\pi\alpha) + 2\epsilon_B^2 L_0(\gamma_1^2 L_0(\pi\alpha))\right). \tag{6.22}$$

Here again, $L_m(x)$ are the Laguerre polynomials.

At a finite bias voltage [Fig. 6.6 (b,c)] the degeneracy of two low energy LLs is lifted and we can observe two distinctively separated butterfly structures for large values of α. For small α (large magnetic field), there is a large overlap of the two butterfly structures and a strong inter-level mixture is expected. In one of the initially degenerate LLs the flatband condition is satisfied for $\alpha \approx 0.35$ [Fig. 6.6 (c)].

6.1.6 Butterflies and interacting electrons

Theoretical studies of the Hofstadter butterfly problem was mostly confined to non-interacting electron systems. There are only a few publications available in the literature that consider the effects of electron-electron interaction on the butterfly spectrum[19]. The main difficulty associated with the interacting systems is the requirement that the system needs to be large enough to capture the fractal nature of the spectrum.

---

[19]Effects of screening on the Hofstadter butterfly, by V. Gudmundsson and R. R. Gerhardts, *Phys. Rev. B* **52**, 16744 (1995); Quantum Hall effect of interacting electrons in a periodic potential, by D. Pfannkuche and A.H. MacDonald, *Phys. Rev. B* **56**, 7100(R) (1997); Effects of electron correlations on the Hofstadter spectrum, by H. Doh and S.H. Salk, *Phys. Rev. B* **57**, 1312 (1998).

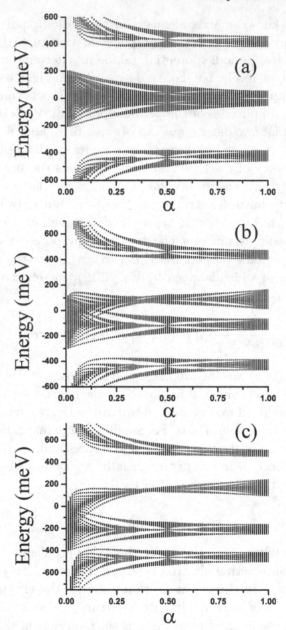

FIGURE 6.6

Single-electron energy spectra of bilayer graphene in a periodic potential and an external magnetic field. The period of the potential is $a_0 = 20$ nm and its amplitude is $V_0 = 100$ meV. The bias voltage is (a) $U = 0$, (b) $U = 200$ meV, and (c) $U = 400$ meV. The energy spectra are shown as a function of the parameter $\alpha = \Phi_0/\Phi$.

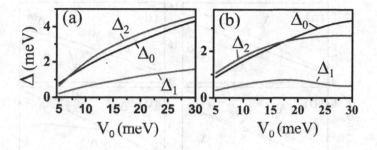

FIGURE 6.7
The band gaps in the $n = 0$, $n = 1$, and $n = 2$ LLs versus the amplitude of the periodic potential, V_0, for interacting systems with half filling of the $n = 0$ Landau level. The band gap Δ_n at LL with index n is defined as the gap between the corresponding bands of Dirac fermions in a magnetic field corresponding to $\alpha = 1/2$. The period of the potential is (a) $a_0 = 20$ nm and (b) $a_0 = 40$ nm.

The influence of the Coulomb interaction between the Dirac fermions, in particular, on the energy gap is highly nontrivial[20]. The graphene LLs with indices $n = 0$, $n = \pm 1$, and $n = \pm 2$ were considered and the gap structure for $\alpha = 1/2$ and $\alpha = 1/3$ with interaction and without interaction were analyzed. The periodic boundary conditions were applied and the size of the system studied was $50a_0 \times 50a_0$, where a_0 is the period of the periodic potential. For $\alpha = 1/2$, the system is expected to have two bands separated by a gap. For noninteracting system the gap is zero at all LLs. Finite electron-electron interactions open gaps for $\alpha = 1/2$, where the magnitude of the gap depends both on the period a_0 and its magnitude V_0. In Fig. 6.7 this dependence is shown for the case when half of the $n = 0$ LL is occupied, i.e., the chemical potential is zero. Strong nonmonotonic dependence of the gaps on the LL index is clearly visible in Fig. 6.7, and as a function of the LL index the gap has a minimum for $n = 1$.

When $\alpha = 1/3$ then even without the interaction, the system has three bands and correspondingly two nonzero gaps in each LL (Fig. 6.8). For a non-interacting system the two gaps $i = 1, 2$ in the LL with index n are labeled as $\Delta_{n,i}^{(0)}$. Due to the symmetry the two gaps in the $n = 0$

---

[20]Gap structure of the Hofstadter system of interacting Dirac fermions in graphene, by V. Apalkov and T. Chakraborty, *Phys. Rev. Lett.* **112**, 176401 (2014).

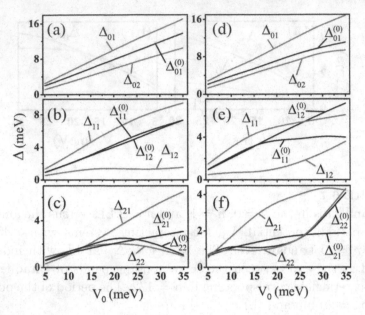

FIGURE 6.8

The band gaps versus V_0 for $n = 0$ (a),(d), $n = 1$ (b),(e), and $n = 2$ (c), (f) LLs. The band gaps are defined as the gaps between the corresponding bands of Dirac fermions in a magnetic field for $\alpha = 1/3$. The gaps are labelled as $\Delta_{ni}^{(0)}$ (noninteracting system) and Δ_{ni} (interacting systems), where n is the LL index and $i = 1$ and 2 correspond to the low-energy and high-energy gaps, respectively. The period of the potential is 20 nm (a)-(c) and 40 nm (d)-(f).

LL are the same, $\Delta_{01}^{(0)} = \Delta_{02}^{(0)}$. In higher LLs ($n = 1$ and 2) the two gaps are different due to the LL mixing introduced by the periodic potential. Then the gaps in the same LL are different, e.g., $\Delta_{11}^{(0)} \neq \Delta_{12}^{(0)}$. The Coulomb interaction modifies the gaps with the general tendency that the lower energy gap is enhanced and the higher one is suppressed. For $n = 0$ the two gaps are no longer equal, $\Delta_{01} \neq \Delta_{02}$. As a function of the amplitude of the periodic potential the gaps have nonmonotonic dependence with local minimum (or maximum) at finite values of V_0. The higher energy gap for $n = 1$, Δ_{12}, is strongly suppressed by the electron-electron interactions (Fig. 6.8).

The enhancement or suppression of the gaps by the electron-electron interactions depend not only on the amplitude of the periodic potential

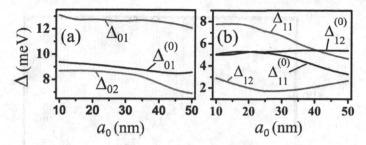

FIGURE 6.9
The band gaps in (a) $n = 0$ and (b) $n = 1$ LLs versus the period a_0 of the periodic potential for non-interacting system and the system with interaction and half filling of the $n = 0$ Landau level for $\alpha = 1/3$. The amplitude of the potential is $V_0 = 25$ meV.

but also on the period of the potential. This dependence is shown in Fig. 6.9 for $\alpha = 1/3$ and amplitude of the potential $V_0 = 25$ meV. The results are shown for the $n = 0$ and $n = 1$ LLs only. The gaps, both for the system with interactions and without interactions, have weak dependence on a_0 for small values of the period, $a_0 \lesssim 25$ nm. For larger values of a_0 there is a strong suppression of the low energy gap, Δ_{11}, in the $n = 1$ LL and higher energy gap, Δ_{02}, in the $n = 0$ LL. In general, the gaps have monotonic dependence on a_0, except the higher energy gap, Δ_{12}, in the $n = 1$ LL, which has a minimum at $a_0 \approx 25$ nm.

In the extreme quantum limit, i.e., in a strong magnetic field and extremely low temperatures, electrons display the celebrated fractional quantum Hall effect (FQHE), which is an unique manifestation of the collective modes of the many-electron system. The effect is driven entirely by the electron correlations resulting in the so-called incompressible states (see Chapter 2 and Chapter 5). The magnetic translation analysis[21] was used to study the fractional quantum Hall effect at the primary filling factor of $\nu = \frac{1}{3}$ (the Laughlin state) in Hofstadter butterflies of Dirac fermions[22]. This work has unveiled a profound effect

[21]Magnetic translation group, by J. Zak, *Phys. Rev.* **134**, 1602 (1964); Bloch electrons in a uniform magnetic field, by E. Brown, *Phys. Rev.*, **133**, 1038 (1964); Fractional quantum Hall effect in a periodic potential, by A. Kol and N. Read, *Phys. Rev. B* **48**, 8890 (1993).

[22]Fractional quantum Hall effect in Hofstadter butterflies of Dirac fermions, A. Ghazaryan, T. Chakraborty, and P. Pietiläinen, *J. Phys.: Condens. Matter* **27**, 185301 (2015).

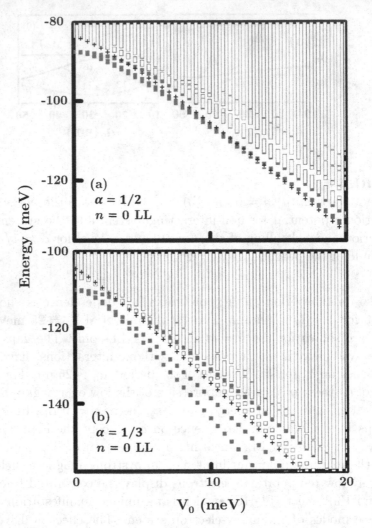

FIGURE 6.10

The low-lying four-electron energy levels versus V_0 for $n = 0$ LL. The results are for (a) $\alpha = 1/2$ and (b) $\alpha = 1/3$. The triplet ground state is shown as filled points and the first excited state which crosses the ground state is plotted with '+'. The other excited states are shown as open points.

of the FQHE states on the butterfly spectrum resulting in a transition from the incompressible FQHE gap to the gap due to the periodic potential alone, as a function of the periodic potential strength. There are

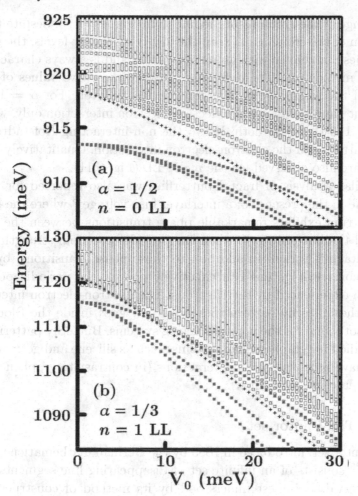

FIGURE 6.11
Same as in Fig. 6.10, but for the $n = 1$ LL.

also crossing of the ground state and low-lying excited states depending on the number of flux quanta per unit cell, that are absent when the periodic potential is turned off.

For $\alpha = 1/2$ and $\alpha = 1/3$ (Fig. 6.10), increasing the periodic potential strength V_0 resulted in a closure of the FQHE gap and the appearance of gaps due to the periodic potential. It was also found that for $\alpha = 1/2$ this results in a change of the ground state and consequently in the change of the ground state momentum. For $\alpha = 1/3$, despite the observation of the crossing between the low-lying energy levels, the ground state does not change with an increase of V_0 and is always characterized by zero momentum. The difference between these two values of α is a result of the origin of the gaps for the energy levels. For $\alpha = 1/2$ the emergent gaps are due to the electron-electron interaction only, whereas for $\alpha = 1/3$ these are both due to the non-interacting Hofstadter butterfly pattern and the electron-electron interaction. Qualitatively similar results are also observed for the $n = 1$ LL (Fig. 6.11).

Similar analysis of fractal butterflies was also reported for bilayer graphene in the presence of a interlayer bias voltage[23] where the butterfly spectrum exhibits remarkable phase transitions between the FQHE gap and the butterfly gap, when the periodic potential strength or the bias voltage is varied. In addition to those phase transitions, by varying the bias voltage one can effectively control the periodic potential strength experienced by the electrons. The electron-electron interaction causes the butterfly pattern to exhibit new gaps inside the Bloch subbands not found for the non-interacting systems. Butterfly patterns were also studied for graphene-like systems, such as silicene and germanene[24], which have large spin-orbit interactions (in contrast to graphene where such interaction is minimal).

6.1.7   The Cantor set

The Cantor set, introduced in 1883 by the German mathematician Georg Cantor[25] consists of an infinite set of disappearing line segments in the unit interval. It is best understood by its method of construction, as shown below for a ternary Cantor set. We start by the following recursive operations on the unit interval $[0, 1]$. The set is generated by removing the open middle third of the unit line segment (step $n = 1$). The length removed is of course, $1/3$. From the remaining two line segments (each

---

[23]Fractal butterflies of chiral fermions in bilayer graphene: Phase transitions and emergent properties, by A. Ghazaryan and T. Chakraborty, *Phys. Rev. B* **92**, 235404 (2015).

[24]Fractal butterflies in buckled graphenelike materials, by V.M. Apalkov and T. Chakraborty, *Phys. Rev. B* **91**, 235447 (2015).

[25]Interestingly, his doctoral thesis at the University of Berlin, which he completed at the age of 22, was entitled, *In re mathematica ars propendi pluris facienda est quam solvendi* (In mathematics, the art of asking questions is more valuable than solving problems). See, *The Foundations of Mathematics 1800 to 1900*, by M.J. Bradley, Chelsea House, N.Y. 2006.

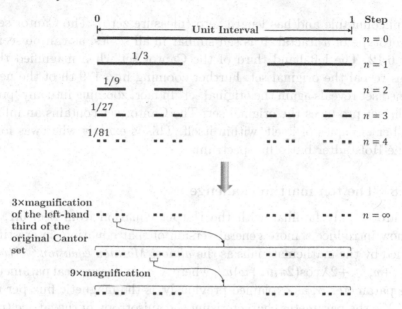

FIGURE 6.12
The construction of the Cantor ternary set.

1/3 in length) we again remove the open middle third ($n = 2$). Now the length removed is $1/3^2$. Again, from the remaining four line segments, each 1/9-th in length, we remove the open middle third ($n = 3$) and the process is repeated infinite times. Only after an infinite number of iterations, we obtain finally the Cantor set (Fig. 6.12).

The set has many interesting (and counter-intuitive) properties[26]. At each step of the construction shown above, the number of intervals doubles and their length decreases by 3. The total intervals removed in the process can be written as a convergent geometric series[27] that results in 1, the same length that we began with! There are as many points in the Cantor set as there are numbers in the real number line. The set consists of the end points (but many other points as well, e.g., the fraction 1/4 is in the Cantor set)[28], and is nowhere dense. Such sets

---

[26] *The Elements of Cantor Sets with Applications*, by R.W. Vallin (John Wiley & Sons, New Jersey 2013).

[27] Total length / measure of intervals removed: $\frac{1}{3} + 2 \cdot \frac{1}{3^2} + 2^2 \cdot \frac{1}{3^3} + \cdots = \sum_{n=0}^{\infty} 2^n \cdot \frac{1}{3^{n+1}} = \cdots = 1$

[28] Because 1/4 remains always one fourth of the way from the left endpoint of each iterated subinterval not removed in the process of constructing the Cantor set. See, for example, Cantor, 1/4, and its family and friends, by I. Mihalia, *The College Mathematics Journal* **33**, 21 (2002).

are uncountable and has length zero (measure zero). The Cantor set is a prototype of a fractal. It is self-similar in all scales, as can be seen in Fig. 6.12. The left-hand third of the Cantor set when magnified three times reveal the original set. Further zooming into 1/9-th of the newly formed set reveals again the original set. In fact, zooming into any 'point' in the set produces the original set. The Cantor set contains an infinite number of copies of itself, within itself. This is exactly what was found in the Hofstadter butterfly spectrum.

6.1.8   The ten-martini challenge

We are already familiar with the Harper equation from Sec. 6.1. Let us now introduce a more general version of that equation, interestingly named by the mathematicians as the *almost Mathieu equation*,[29] $\mathcal{H}u_n = u_{n+1}+u_{n-1}+2\lambda\cos(2\pi n\alpha+\varphi)u_n$, where λ, α, and φ are real parameters. The parameter α, as explained previously, is the magnetic flux per unit cell, λ is the parameter characterizing the anisotropy of the lattice (ratio between the length of a unit cell in two directions), and φ is the phase shift per lattice constant in the y direction with a wave number $k_y = \varphi/a_0$. For the isotropic case of $\lambda = 1$, the almost Mathieu (AM) equation is the Harper equation.

There has been a longstanding theoretical challenge for the mathematicians to explain the self-similar structure of the solution of the almost Mathieu (AM) equation, as numerically demonstrated by the Hofstadter butterfly, and the hierarchical structure of the eigenfunctions, predicted by Azbel[30] in 1964. As explained above, the butterfly spectrum crucially depends on the Diophantine properties of α. For instance, if $\alpha = p/q$ is a rational number, p, q being positive integers, then the AM operator is periodic with period q. Consequently, the spectrum consists of q continuous areas called *bands*, separated by gaps. However, the situation is very different for irrational α (the spectrum is then independent of φ), which is approximated by a set of rational approximants $\alpha_n = p_n/q_n$ with $n \to \infty$. Here α is the root of a polynomial with integer coefficients (i.e., it solves a diophantine equation). A prime example is the golden mean $\alpha_G = \left(1 + \sqrt{5}\right)/2$, which solves the equation $x^2 = x + 1$. This number is irrational and has the worst approximation

---

[29]Named as such due to it being a discrete version of the Mathieu equation: $-u''(x) + 2\lambda\cos(2x)u(x) = Eu(x)$.

[30]Energy spectrum of a conduction electron in a magnetic field, by M.Y. Azbel, *Sov. Phys. JETP* **19**, 634 (1964).

by a continued fraction:

$$\alpha_G = 1 + \cfrac{1}{1 + \cfrac{1}{1 + \cfrac{1}{1 + \cdots}}}$$

because the n-th remainder of the fraction does not converge to zero, it is just α_G itself! In general, the diophantine numbers are those irrationals that has no good (i.e., fast converging) approximation, but they are countable many, as they are all roots of diophantine polynomials. The liouvillian numbers, on the other hand, are very good approximable by continued fractions and those form already an uncountable set. For irrational α the spectrum is fractal. A word of caution is in order here: the Diophantine numbers as defined are only a countable subset of those, i.e., the algebraic numbers connected with the Diophantine equations (as the golden mean). But the set of Diophantine numbers is actually much larger, in fact, uncountable and according to the Appendix (p. 221) even so large that the irrational minus the Diophantine numbers form a set of Lebesgue measure zero (this latter set is still uncountable, and contains the Liouville numbers).

In 1981, during his talk in a session of American Mathematical Society, mathematician Mark Kac offered ten martinis to anyone who can prove that the AM operator has the Cantor spectra for any irrational α. It was then popularly dubbed as the 'Ten-Martini Problem'. The challenge was to prove that for *every* irrational α and every choice of the coupling constant $\lambda > 0$, the spectrum is a Cantor set. Facing this challenge, a surge of activities ensued and the problem was partially solved by various authors under various conditions[31]. It was generally established that the Hamiltonian has the Cantor spectrum for almost every (a.e.) irrational α, either being well approximated by rationals (i.e., a liouville number) or those that are poorly approximated by rationals (the diophantine region)[32]. Apparently, in solving the ten martini problem, it matters how precisely these two regions are defined (see Box 6.6). There still remained a gap between those two cases, where α is neither diophantine nor liouvillian and that gap required a different strategy.

---

[31] *Almost Everything About the Almost Mathieu Operator*, by Y. Last, XIth International Congress of Mathematical Physics (Paris, 1994), Internat. Press, Cambridge, MA, 1995, p. 366.

[32] Almost periodic Schrödinger operators: A review, by B. Simon, *Adv. Appl. Math.* **3**, 463 (1982).

Box 6.6 Liouville or Diophantine?

In solving the ten-martini problem, it is clear that the arithmetic of α plays the crucial role. To decide whether α should be considered Liouville or Diophantine for the ten-Martini problem, Avila and Jitomirskaya introduced a parameter[33]

$$\beta(\alpha) = \limsup_{n \to \infty} \frac{\ln q_{n+1}}{q_n}$$

where p_n/q_n is the n-th rational approximant of α obtained from the continued fraction expansion of α. As β increases, the Diophantine method becomes less and less efficient, until it does not work any longer. For the Liouville method, the opposite is true. The two methods however, do not meet in the middle, and these authors found a range, $\beta \leq |\ln \lambda| \leq 2\beta$ where a new approach was required to solve the problem of proving the existence of the Cantor set.

In the Diophantine case, an important breakthrough was the work of Puig who showed that for $\lambda = 0, \pm 2$, the spectrum is a Cantor set for α satisfying a diophantine condition[34]. Finally, Avila and Jitomirskaya[35] completed the proof on the Liouville side and then they also established that the spectrum is a Cantor set for all real $\lambda \neq 0$ and irrational α. The analysis of Avila and Jitomirskaya in solving this important and interesting problem is rather involved and is described briefly in the Appendix by S. Jitomirskaya.

Although the problem was finally solved, no martini was however, 'shaken' or 'stirred' by the challenger because Kac died in 1984, long before all these wonderful developments took place.

---

[33]Solving the Ten Martini Problem, by A. Avila and S. Jitomirskaya, *Lect. Notes Phys.* **690**, 5 (2006).

[34]Cantor spectrum for the almost Mathieu operator, by J. Puig, *Commun. Math. Phys.* **244**, 297 (2004).

[35]The Ten Martini Problem, by A. Avila and S. Jitomirskaya, *Ann. Math.* **170**, 303 (2009).

6.2 Maxwell's demon in the nanoworld

Perhaps no other law of physics evokes as much conflicting and often confusing opinions as the second law of thermodynamics[36]. Ever since it made its debut around 1850, physicists, engineers, mathematicians and philosophers have joined in the collective onslaught of the second law. Consider these: The physicist and philosopher Percey W. Bridgman wrote in 1943 that there are almost as many formulations of the second law as there have been discussions of it[37]! The historian of science and mathematician Clifford A. Truesdell characterized thermodynamics as a 'dismal swamp of obscurity'[38]. There are many other such examples in the literature. Then, at the other end, Lieb and Yngvason[39] began their article with unusual praise of the second law: 'The second law of thermodynamics is, without a doubt, one of the most perfect laws in physics. Any 'reproducible' violation of it, however small, would bring the discoverer great riches as well a trip to Stockholm. The world's energy problems would be solved at one stroke ...'. Clearly, that trip to Stockholm is not meant entirely for sight seeing!

Maxwell described the law as follows[40]:'One of the best established facts in thermodynamics is that it is impossible in a system enclosed in an envelope which permits neither change of volume nor passage of heat, and in which both the temperature and the pressure are everywhere the same, to produce any inequality of temperature or pressure without the expenditure of work. This is the second law of thermodynamics, and it is undoubtedly true so long as we can deal with bodies only in mass and have no power of perceiving or handling the separate molecules of which they are made up.' Spontaneous heat transfer from hot to cold bodies is an irreversible process and there is an increase in 'entropy' for any system undergoing an irreversible process.

---

[36] The Second Law of Thermodynamics: Foundations and Status, by D.P. Sheehan, *Found. Phys.* **37**, 1653 (2007).

[37] *The Nature of Thermodynamics*, by P.W. Bridgman, Harvard University Press, Cambridge, 1943.

[38] *The Tragicomical History of Thermodynamics 1822 - 1854*, by C.A. Truesdell, III (Springer-Verlag, N.Y. 1980).

[39] The physics and mathematics of the second law of thermodynamics, by E.H. Lieb and J. Yngvason, *Phys. Rep.* **310**, 1 (1999).

[40] *Theory of Heat*, by J.C. Maxwell, (Longman, London, 1871).

Entropy is in fact, a measure of the disorder of a system. It also describes how much energy is not available to do work. The more disordered a system is and higher the entropy, the less energy of a system is available to do work. A very amusing example of entropy and the ensuing change in disorder can be found in a book that describes it as follows[41]: 'Clausius's definition of the change in entropy is that of sneezing in a busy street or in a quiet library. A quiet library is the metaphor for a system at low temperature, with little disorderly thermal motion. A sneeze corresponds to the transfer of energy as heat. In a quiet library a sudden sneeze is highly disruptive: there is a big increase in disorder, a large increase in entropy. On the other hand, a busy street is a metaphor for a system at high temperature, with a lot of thermal motion. Now the same sneeze will introduce relatively little additional disorder: there is only a small increase in entropy. Thus, in each case it is plausible that a change in entropy should be inversely proportional to some power of the temperature (the first power, T itself, as it happens; not T^2 or anything more complicated), with the greater change in entropy occurring the lower the temperature. In each case, the additional disorder is proportional to the magnitude of the sneeze (the quantity of energy transferred as heat) or some power of that quantity (the first power, as it happens).'

The fact that heat only flows from hot to cold systems (without any external input) and never the reverse is also associated with the thermodynamic time asymmetry or the 'arrow of time' that distinguishes past from the future[42]. However, that association of the second law with arrow of time has not been universally agreed upon[43]. There are also other profound implications of the law which indicates that, as the entropy is forever increasing, the universe, according to Kelvin, is bound to come to a state of eternal rest. The entire universe then has a constant temperature at all points in space and no work can be done – it is the

---

[41]Four laws that drive the universe, by P. Atkins, Oxford University Press (Oxford, 2007).

[42]'The past is that which is no more; the future is that which is not yet. And if the present were perpetually present, there would be no longer any time, but only eternity. For the present to belong to time it must pass. Hence time only exists because it tends to not-being', by Herman Hausheer in St. Augustine's Conception of Time, *The Philosophical Review* **46**, 503 (1937); Direction of time, edited by S. Albeverio and P. Blanchard (Springer, Heidelberg 2014).

[43]Bluff your way in the second law of thermodynamics, by J. Uffink, *Stud. Hist. Phil. Mod. Phys.* **32**, 305 (2001); What is 'the problem of the direction of time'? by C. Callender, *Philosophy of Science* **64**, S223 (1997).

so-called *heat death of the universe*[44], as envisaged by H.G. Wells in the closing pages of the science fiction novella 'The Time Machine' (1895).

As an illustration of the implications of the second law, let us consider a container separated into two compartments by a trap door and filled with the same amount of *ideal* gas (no interaction among the gas molecules) and at the same temperature. If we open the door then the particles of both compartments will fill the entire space in the container. Interestingly, while we can be fairly confident that the possibility of all the particles spontaneously moving at some point in time into one compartment, while the other compartment remains empty, is close to zero, it is actually not exactly zero. In fact, one can estimate the time it would take to come to that situation, which will be extremely long, and might even be close to the age of the universe. We can therefore safely dismiss that state ever happening. Nonetheless, it is not strictly forbidden by the laws of physics. In this respect, the second law is quite unique: Unlike every other known fundamental laws of physics this law forbids the anti-thermodynamic process with only high probability, but not absolutely.

6.2.1 A sorting demon – the anti-thermodynamic agent

Let us again go back to the quote above by Maxwell, where he expressed our inability to 'perceive or handle separate molecules'. But what would happen if we can somehow achieve that imperceptible? Again in Maxwell's own words: 'if we conceive a being whose faculties are so sharpened that he can follow every molecule in its course, such a being, whose attributes are still as essentially finite as our own, would be able to do what is at present impossible to us.' That 'neat-finger being' of Maxwell was later christened by William Thomson (Lord Kelvin) as the 'sorting demon of Maxwell'[45] and Maxwell's demon was born.

Unlike Mephistopheles the famous demon in the literature, Maxwell's demon lacks any description of his appearance, other than that he is super-intelligent. He was described by Kelvin as a 'being with no preternatural qualities, and differs from real living animals only in extreme

---

[44]It should be pointed out that, following the big-bang theory, a continually expanding universe never reaches true thermodynamic equilibrium and therefore never reaches a constant temperature. See, for example, A dying universe: the long-term fate and evolution of astrophysical objects, by F.C. Adams and G. Laughlin, *Rev. Mod. Phys.* **69**, 337 (1997).

[45]The sorting demon of Maxwell, by William Thomson, *Nature* **20**, 126 (1879).

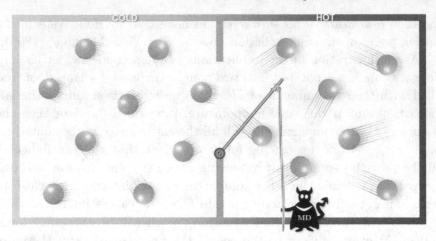

FIGURE 6.13
Maxwell's sorting demon at work.

smallness and agility. He can at pleasure stop, or strike, or push, or pull any single atom of matter, and so moderate its natural course of motion. Endowed ideally with arms and hands and fingers - two hands and ten fingers suffice - he can do as much for atoms as a pianoforte player can do for the keys of the piano - just a little more, he can push or pull each atom in any direction.' Although the Enlightenment has taught us about the power of reason and to dispense with the supernatural, we need not be concerned to welcome this demon in our midst because, as Kelvin pointed out, 'The word 'demon,' which originally in Greek meant a supernatural being, has never been properly used to signify a real or ideal personification of malignity.' The demons in science are not associated with the forces of darkness, but are agents with superhuman abilities. They are called upon to provide invaluable insights about the constitution of the physical world[46].

The primary job of Maxwell's sorting demon can be simply described as demonstrated in Fig. 6.13. The system is again a container, divided into two chambers by a partition equipped with a frictionless door. Although the temperature of the gas is uniform, the velocities of the gas molecules need not be, since the temperature is the average kinetic energy of the molecules. The job of the demon, who could follow the

---

[46] *The Demons of Science* by F. Weinert (Springer, Heidelberg 2016).

movements of individual molecules, is to open and close the door so as to allow the swifter molecules to pass from left to right, and only the slower ones to pass from right to left. With time, according to Maxwell the demon will thus, without expenditure of work, raise the temperature of the right chamber and lower that of the left, in contradiction of the second law of thermodynamics. If it is broken, the second law can no longer enjoy the deterministic validity. It must be a statistical law, which holds with high probability for a many-particle system.

Irreversible processes are, for all practical purposes, inherently wasteful. In fact, Maxwell's own idea of introducing the demon was to show that 'energy need not be dizzypated as in the present wasteful world.'[47]. If Maxwell's insight regarding the demon is correct, we will then have a limitless supply of energy. A Maxwellian demon could be installed in every windows, in every engines that powers the airplanes, automobiles, power plants etc., without any loss of heat (Fig. 6.14) in the surroundings (global warming will be a thing of the past). Additionally, the heat death, the dire consequence brought upon the universe by the inexorable second law could also be averted in this process. But where can we find such an important and essential demon? After nearly 150 years since he was first conjured, the 'restless and lovable poltergeist'[48] is now back with mind-blowing ideas. As we have discussed in the earlier chapters of this book, in nanoscale devices, e.g., quantum dots, quantum rings, etc., control over even a single electron is possible. The question then is, could these systems be the workplace of the sorting demon?

6.2.2 Anthropomorphism of the benevolent demon

For law abiding physicists, breaking the second law is perhaps not so appealing. Therefore, it is no wonder that since the mid-twentieth century, various proposals have been put forward to demonstrate that the demon must fail, while the second law must prevail. In these proposals, the demon is itself considered to be a physical system. The object system (demon's workplace) and the demon form a larger system that is governed by the second law. It has been argued (Szilard's principle)[49]

---

[47]Maxwell's demon, by E. Daub, *Studies in History and Philosophy of Science* **1**, 213 (1970).

[48]*The Enigma of Time*, by P.T. Landsberg (Adam Hilger Ltd., Bristol 1982).

[49]On the decrease of entropy in a thermodynamic system by the intervention of intelligent beings, by L. Szilard, *Z. Phys* **53**, 840 (1929).

FIGURE 6.14
Demon at work: Keeping the tea warm by removing the cold atoms.

that the *information* gathered in demon's memory (considered itself a physical system) causes an increase in the physical entropy. That entropy increase precisely negates the entropy decrease due to demon's sorting process. Therefore, the second law that governs the combined system is saved by considering the demon as one of us. In another twist in this line of argument, several researchers, most notably, Landauer[50] have claimed that it is not the information gathering in demon's brain that causes the necessary entropy increase, but it is the information *erasure*[51] in demon's memory, required by the demon to operate cyclically, that is the source of entropy increase of the total system. As the Demon needs to remember the measurements he makes, at some point he will also need to clear out that space to make room for more data. The destruction of that information results in entropy increase (Landauer's principle). Once again, that increase would be sufficient to protect the

---

[50]Irreversibility and heat generation in the computing process, by R. Landauer, *IBM J. Res. Dev.* **3**, 183 (1961); The thermodynamics of computation – a review, by C.H. Bennett, *Int. J. Theor. Phys.* **5**, 905 (1982).

[51]Information erasure, by B. Piechocinska, *Phys. Rev. A* **61**, 062314 (2000).

second law. The concept of thermodynamic principle of information processing initiated by these authors has become widely popular[52], and the fact that the erasure of information is inevitably accompanied by the generation of heat, has been widely accepted by the community[53].

Interestingly, there has also been rather vehement opposition to these form of 'exorcising' the demon by means of the Szilard's principle and the Landauer principle. Both these principles were termed 'dubious', and at best 'interesting speculations in need of precise grounding'[54]. Maxwell's demon need not always require computation or information processing, as this author posited, in doing its job. Another author surmised that, there are no basic principles and fundamental laws of physics able to absolutely forbid the violation of the second law[55]. According to this author, 'we have not yet any truly cogent argument (known fundamental physical laws) to exclude its possible macroscopic violation'. There are other authors who would also like to see that the demon succeeds in its pursuit and is observed or constructed in the future[56]. Perhaps, there is some hope for demon's survival in some cleverly designed systems.

6.2.3   Demon in quantum dots

A very interesting experimental setup to create a demon in a quantum dot-like system was reported[57] in 2015. The device is made up of two single-electron boxes: a 'system' box where an electron is coaxed to enter or exit, and a 'demon' box that is Coulomb coupled to the system box. The system box consists of a small metal island connected to two

---

[52]Information: From Maxwell's demon to Landauer's erasure by E. Lutz and S. Ciliberto, *Phys. Today* **68**, 30 (2015).

[53]The physics of forgetting: Landauer's erasure principle and information theory, by M.B. Plenio and V. Vitelly, *Contemp. Phys.* **42**, 25 (2001).

[54]Maxwell's demon does not compute, by J.D. Norton, in *Physical Perspectives on Computation, Computational Perspectives on Physics*, edited by M.E. Cuffaro and S.C. Fletcher (Cambridge University Press, Cambridge 2018); Eaters of the lotus: Landauer's principle and the return of Maxwell's demon, by J.D. Norton, *Studies in the History and Philosophy of Modern Physics* **36**, 375 (2005).

[55]The peculiar status of the second law of thermodynamics and the quest for its violation, by G. D'Abramo, *Studies in the History and Philosophy of Modern Physics* **43**, 226 (2012).

[56]The road to Maxwell's demon, by M. Hemmo and O.R. Shenker (Cambridge University Press, Cambridge 2012); Maxwell's demon, by M. Hemmo and O. Shenker, *J. Phil.* **107**, 389 (2010).

[57]On-chip Maxwell's demon as an information-powered refrigerator, by J.V. Koski, et al., *Phys. Rev. Lett.* **115**, 260602 (2015).

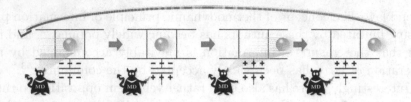

FIGURE 6.15
Maxwell's demon in electronic systems. The operation principle is described in the text.

metal leads via superconducting aluminum junction. The two junctions allow electron transport by tunneling onto and off the island. The demon box has a similar electronic structure. The entire structure is such that Coulomb blockade ensures only one or zero electrons reside in the system island (Fig. 6.15).

When an electron tunnels into the system box, the resultant change in the charge triggers the demon to react and trap the electron by applying a positive charge. If an electron leaves the system, the demon repels it by applying a negative charge (Fig. 6.15). Electrons tunnel against the potential induced by the demon which lets the system to cool down. In fact, the demon is in the details! The energetics of the whole process is shown in Fig. 6.16. When an electron enters the system island from a source electrode, an electron tunnels out of the demon island due to the mutual Coulomb repulsion between the two electrons. Similarly, when an electron tunnels out of the system and enters the drain electrode, an electron tunnels back to the demon island, as it is attracted by the overall positive charge. The cycle is then repeated. As reported by these authors, the demon's action results in a drop in temperature of the system, but the temperature of the demon increases as well. These authors claim that the device presented here demonstrates the transfer of information from the system to the demon, and the amount of heat generation in the demon corresponding to the rate of information transfer. Finally, it is worth mentioning that there has been a theoretical proposal reported on creating the demon in coupled quantum dots[58].

---

[58]Thermodynamics of a physical model implementing a Maxwell demon, by P. Strasberg, et al., *Phys. Rev. Lett.* **110**, 040601 (2013).

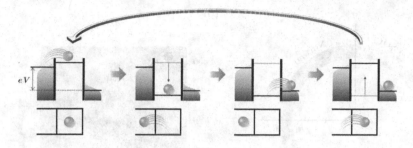

FIGURE 6.16
Energetics of the operation described in Fig. 6.15.

6.2.4 Spin demons in quantum rings

Demon's workplace might as well be a quantum ring. A nanoscale system of 'non-reciprocal quantum ring', where one arm of the ring contains the Rashba spin-orbit coupling (RSOC) [see Sec. 3.5], while the other arm is normal, i.e., without the RSOC [Fig. 6.17 (a)] exhibits all the indications of being the demon's abode[59]. Here the electrons can be channelled through the two arms according to their spin states by switching on and off (or reverse) an external magnetic field. Most significantly, this leads to different kinetic energies and effective spin temperatures in the two arms, not unlike the expected signature of Maxwell's demon. In this system, the demon neither infuses energy or create entropy of its own, nor does it process information.

In the absence of the magnetic field, the electron charge is distributed all over the ring with a slight shift into the arm of the ring where the RSOC is present [Fig. 6.17 (b)]. In this case, the average spin in the ground state is almost zero because the RSOC mixes the spin-up and spin-down states. However, when the magnetic field is present, the electron moves to the area of the ring where the RSOC is absent [Fig. 6.18 (a)-(b)]. The ground state electron density indicates that, with an increase of the magnetic field the value of the average spin for the ground state is positive and increases with the magnetic field. Therefore, the electron is confined in the non-Rashba (right) arm of the ring. If the magnetic field direction is reversed, the average electron spin in the ground state is negative and now the electron is confined in the Rashba (left)

---

[59]Seeking Maxwell's demon in a non-reciprocal quantum ring, by A. Manaselyan, W. Luo, D. Braak, and T. Chakraborty, *Sci. Rep.* **9**, 9244 (2019).

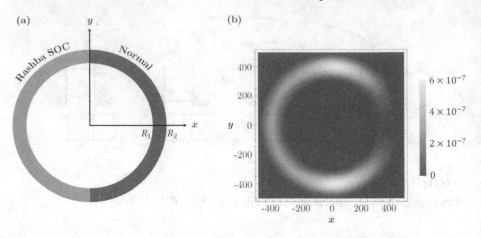

FIGURE 6.17
(a) A non-reciprocal quantum ring. (b) Ground state electron density in a quantum ring in the absence of an external magnetic field.

arm of the ring [Fig. 6.18 (c)-(d)]. Clearly, the external magnetic field acts here as the 'trap door' for the demon to sort the electrons according to their spin.

Maxwell inferred the temperature difference in the two chambers of his gedankenexperiment from different expectation values of the kinetic energy (fast molecules in one chamber and slow molecules in the other) of a classical ideal gas. In the present quantum system, the corresponding quantity can also be evaluated. The kinetic energies of the two arms differ greatly in the non-reciprocal QR [Fig. 6.19 (a)-(b)]. The particles with high and low kinetic energies can therefore be separated into two distinct spatial regions. Hence it is the simplest manifestation of a demon-like process without the interference of an intelligent being. The magnetic field could be turned on or off or reversed adiabatically so that no entropy is generated during the process. To asses the action of our demon, we have introduced a canonical temperature associated with the spin degree of freedom and the associated 'spin entropy' (see Footnote 59 for details). The result is that spin temperatures in the two arms of the non-reciprocal ring are completely different [Fig. 6.19 (c)-(d)]. Although these temperatures cannot be associated with the total system, which remains always at $T = 0$, the difference between the left and right arm temperatures can be expressed as different temperatures of a 'spin gas' in the spirit of Maxwell. As for the spin entropy [Fig. 6.19 (e)], we

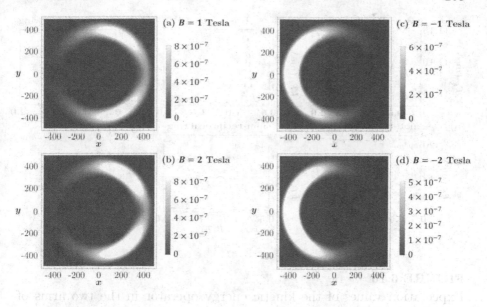

FIGURE 6.18
(a) Ground state electron density for (a) B =1 Tesla, and (b) B = 2 Tesla, (c) B = -1 Tesla and (d) B = -2 Tesla.

notice a sharp drop in the entropy of the 'cool' arm (with the SOC) and a rather flat curve in the 'hot' arm (without the SOC). Both temperature and entropy of the spin subsystem indicate the sorting action of the ring but not a deviation from equilibrium of the total system. In this system, the spin degree of freedom is entangled with the spatial degree of freedom through the Rashba coupling. The spin, acting as a quantum marker, is 'measured' by the magnetic field without recording the information and causes the sorting in both halves of the ring. The non-reciprocal QR is clearly a manifestation of demon's workplace: the electron sorting in different arms of the ring according to their spin leads to different kinetic energies and spin temperatures in the two arms of the ring. This unusual sorting by the demon takes place through the spin, which is a pure quantum character and does not have any classical counterpart.

A fundamental problem of quantum physics is the connection to the classical world, nowadays called 'decoherence'. It describes how one transitions from a deterministic to a stochastic description of the quantum phenomena. A standard way to do that is using the 'golden rule' of

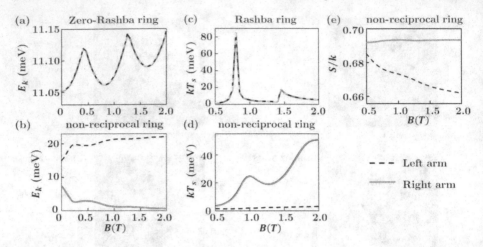

FIGURE 6.19
Expectation values of the kinetic energy operator in the two arms of a (a) Zero-Rashba ring, (b) a non-reciprocal ring, as a function of the magnetic field. The spin temperatures in the two arms of (c) Rashba ring, and (d) non-reciprocal ring as a function of the magnetic field. (e) Spin entropy in the two arms of a non-reciprocal ring versus the magnetic field.

perturbation theory named after E. Fermi (although it has been introduced first by P.A.M. Dirac). In an interesting paper[60] the authors demonstrated that if one employs the golden rule to compute the behavior of a so-called 'chiral' waveguide for photons, the waveguide acts as a kind of demon capable of sorting the photons between two cavities. No information is processed by this demon, just like in the non-reciprocal ring described above.

Normal demons have been banished from our contemporary life. Those small horned creatures or frightening goat-like monsters are too uncouth to be allowed to remain within our sophisticated world. However, the demons in science are not merely the 'auxiliary midwives who help scientists deliver knowledge'[61] but they should be regarded as similar to more established philosophical concepts. Given the profound

────────────────

[60]Fermi's golden rule and the second law of thermodynamics, by D. Braak and J. Mannhart, *Found. Phys.* **50**, 1509 (2020).

[61]*A Shadow History of Demons in Science*, by J. Canales (Princeton University Press, Princeton 2020).

implications of the ground-breaking capabilities of Maxwell's demon, we should perhaps suppress the present urge to exorcise the 'very observant and neat fingered being' and allow him see the light of day.

6.3 Nanoscale physics of DNA

In 1944, Erwin Schrödinger, Nobel laureate and renowned for his ground-breaking contributions to the fundamentals of quantum mechanics, pondered in a book[62] *What is Life?* about the problems of genetics and its relation to the known laws of physics. Using the laws of thermodynamics as a backdrop, he asked how the living systems seem to evade the second law of thermodynamics by remaining highly ordered and even equip the next generation with that trait. His answer was that life can resist the state of maximum entropy, i.e., death, by eating, drinking, breathing, etc. (viz. via metabolism) or as he explained, by homeostatically maintaining 'negative entropy' in an open system. He also stated in that book that life is distinguished by a 'code-script' that directs cellular organization and heredity. That code-script is known today as DNA. Schrödinger proposed that the genetic information is arranged in an 'aperiodic crystal', that is a structure without a periodic arrangement of atoms, which he believed to be the 'material carrier' of life: 'We believe a gene – or perhaps the whole chromosome fibre – to be an aperiodic solid'. In fact, he postulated that a few building blocks could generate an enormous number of combinations. Thus encoding information that somehow guides the development of the organism. Francis Crick, who with James Watson discovered the structure of DNA and how DNA's double helix encodes genes, wrote in 1953 to Schrödinger that he and Watson had 'both been influenced by your little book'.

6.3.1 DNA – Nature's nanoscale code-script

Tucked neatly inside the nucleus of every cell in almost every living being is the molecule of life. DNA (deoxyribonucleic acid) is responsible for preserving, copying and transmitting information that are necessary

---

[62] *What is Life? with Mind and Matter and Autobiographical Sketches*, by Erwin Schrödinger, Cambridge University Press 1944.

for living being to grow and function. It is often referred to as the body's hereditary material as DNA is replicated and transmitted from parents to the offspring. DNA was discovered in 1869 by Johannes Friedrich Miescher in Tübingen, Germany, who called it 'nuclein', as it was isolated from the nucleus of white blood cells[63], long 75 years before the discovery of the DNA structure by Watson and Crick[64].

There are countless sources available for detailed description of the DNA structure and function[65]. Here we present very succinctly the basics of the DNA double helix structure[66]. In the DNA duplex, each nucleotide [Fig. 6.20 (a)] contains three components: a heterocyclic base, a deoxyribose sugar (pentose), and a phosphate (phosphoric acid). The sugar and phosphate of the successive nucleotide units along each chain are connected in an alternating sequence and form the backbone of the chain. The base of each nucleotide attaches to the sugar on one side and to its counterpart base from the other chain on the other side. The two chains are held together through pairing of their bases by hydrogen bonds. There are four kinds of bases, two purine derivatives, guanine (G) and adenine (A), and two pyrimidine derivatives, cytosine (C) and thymine (T). The pairing occurs only between G and C by three hydrogen bonds or between A and T by two hydrogen bonds, i.e., there are only two kinds of canonical base pairs, (G:C) and (A:T) [Fig. 6.20 (b)]. Along each backbone, the phosphate connects the carbon 5 of one sugar with the carbon 3 of the next sugar.

The secondary structure of DNA is a double helix with the duplex nucleotide strands twisted around each other. The two strands of the nucleotide polymer in a DNA are oriented in opposite directions, one from carbon 5 to 3 and the other from carbon 3 to 5. The

---

[63]Friedrich Miescher and the discovery of DNA, by Ralf Dahm, *Developmental Biology* **278**, 274 (2005); Serendipity and the discovery of DNA, by A. de Sojo, et al., *Found. Sci.* **19**, 387 (2014).

[64]Molecular structure of nucleic acids, by J.D. Watson and F.H.C. Crick, *Nature* **171**, 737 (1953); The discovery of the DNA double helix, by A. Klug, *J. Mol. Biol.* **335**, 3 (2004).

[65]*The Eighth Day of Creation: Makers of the Revolution in Biology*, by H.F. Judson, Cold Spring Harbor Laboratory Press, New York 1996; *DNA Structure and Function*, by R.R. Sinden, Academic Press, London 1994; *DNA Science: A First Course*, by D.A. Micklos, et al., Cold Spring Harbor Laboratory Press, New York 2003.

[66]*A Wonderful Account of DNA for Non-experts can be Found in, Understanding DNA: The Molecule and How it Works*, by C.R. Calladine, H.R. Drew, B.F. Luisi, and A.A. Travers, Elsevier (2004).

antiparallel orientation helps to align the hydrogen bond donors and acceptors. Along the double helix, the two strands of the backbone wrap around the stacked base-pair layers. There are three classes of structures, called the B, A, and the Z forms. The form of the B-DNA commonly exists in living beings where the environment is humid. Its helix is about 2 nm in diameter with a vertical distance of about 0.34 nm between layers of the base pairs and about 10 base pairs for each complete turn of the helix [Fig. 6.20 (c)]. Double stranded DNA is one of the most highly charged polymers with a bare charge density of one charge per 0.17 nm[67].

The significance of the discovery of the double helix structure of DNA is so profound in biology that it has been compared with the discovery of the atomic nucleus in physics[68]. Understanding the structure of the atom ushered in the quantum physics, while the understanding of the structure of DNA brought the advent of molecular biology. The blueprints for making proteins (the complex molecules that do most of the work in our bodies) are stretches of DNA called genes. There the instructions are written in four letter codes: A, T, G, and C. To make a copy of itself, DNA simply unzips along its length forming two half-ladders that are reverse images of each other. Each half then rebuilds itself by matching the corresponding bases to form the base pairs. It is interesting to note that all the trillions of cells that compose our body[69], contain the same 3 billion DNA base pairs, our entire genetic makeup. DNA replicates itself at each division of the cell, an essential process when a cell duplicates all of its contents, including its chromosomes, and splits to form two identical daughter cells[70]. In this process, every cell carries information about the entire organism. This is like each stone of a building containing a blueprint of the entire edifice. That way, the information will never be lost. In fact, the building can be reconstructed even from a single stone. Rather amusingly, the 'homunculi' in human sperm, as depicted

---

[67]DNA-inspired electrostatics, by W.M. Gelbart, et al., *Phys. Today* **53**, 38 (2000).

[68]*Unraveling DNA: The Most Important Molecule of Life*, by M.D. Frank-Kamenetskii, Basic Books (1997).

[69]An estimation of the number of cells in the human body, by E. Bianconi, et al., *Annals of Human Biology* **40**, 463 (2013).

[70]A very interesting brief introduction to cell biology can be found in: Omnis cellula e cellula revisited: cell biology as the foundation of pathology, by N.A. Wright and R. Poulsom, *J. Pathol.* **226**, 145 (2012).

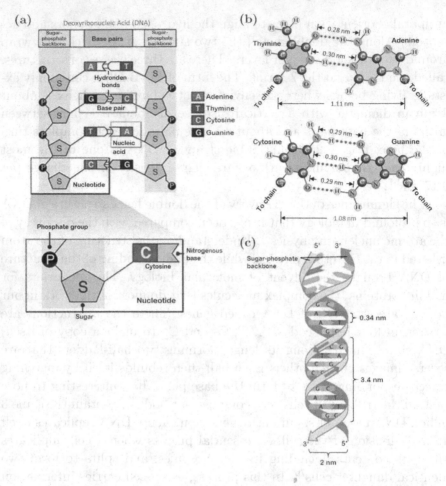

FIGURE 6.20
(a) The nucleotides, (b) the canonical base pairs and (c) the double helix
structure of DNA.

by the mathematician and physicist Nicholaas Hartsoecker[71] in 1695 has
turned out to be a compactified double helix of DNA carrying the genetic
information!

An interesting estimate about the possible combinations of four bases
in each copy of the genome in human DNA: we have about 6×10^9 base
pairs in our DNA. There are four types of nucleotide bases in DNA.

---

[71]Strange Tales of Small Men: Homunculi in Reproduction by C. Pinto-Correia,
Perspectives in Biology and Medicine, **42**, 225 (1999).

Therefore, there are approximately $4^{6,000,000,000}$ possible combinations of bases in human DNA. In this vast design space the genome could plan all kinds of as yet unforeseen life forms. After all, Delbrück's alleged 'stupid molecule'[72], that was regarded as too simple to be the leading molecule of life, DNA is in fact, more than capable of handling the complexity of present (and past) life forms and then some!

6.3.2 DNA electronics

DNA, Nature's most remarkable molecule has several unique properties that has made this molecule a prime candidate for nano-electronic materials[73]. These include, molecular recognition and the ability to self-assemble via the complementarity of the base sequences on the two strands and the double-helical nature of the polymer. The base sequence complementarity, with adenine complementary to thymine and guanine complementary to cytosine means that DNA can be integrated error-free in current semiconductor technology. Incidentally, DNA can detect information about its own integrity that can trigger its eventual repair mechanism, that might also be an useful property for DNA electronics. Self-assembly of molecular building blocks into well-structured systems will allow one to avoid physical manipulation during fabrication of nanoscale devices. However, most importantly, does DNA conduct electricity? There is an wide range of studies on this question and will not be reviewed here[74], but the answer is as yet surprisingly unclear. It seems that DNA with random base pairs is actually an insulator. However, it is a wide-gap semiconductor for a periodic DNA, a 10.4 nm long, repetitive synthetic double-stranded DNA sequence connected to two metal electrodes[75]. Other measurements of poly(dG)-poly(dC) and poly(dA)-poly(dT) showed semiconductor-like behavior of the p-type and n-type, respectively[76].

While pristine DNA is not a good conductor, its conductivity was found to be greatly enhanced by suitably changing the environment, in

---

[72] *One Plus One Equals One: Symbiosis and the evolution of Complex Life*, by John Archibald (Oxford University Press, 2014).

[73] DNA electronics, by M. Taniguchi and T. Kawai, *Physica E* **33**, 1 (2006).

[74] See for example, *Charge Migration in DNA*, edited by T. Chakraborty (Springer, New York 2007).

[75] Direct measurement of electrical transport through DNA molecules, by D. Porath, et al., *Nature* **403**, 635 (2000).

[76] Electrical conduction through Poly(dA)-Poly(dT) and Poly(dG)-Poly(dC) Molecules, by K.-H. Yoo, et al., *Phys. Rev. Lett.* **87**, 198102 (2001).

particular, the humidity is recognized as an important factor controlling the DNA conductivity.

6.3.3 Humidity assisted conduction

Several experiments on DNA conductance have exhibited an exponential increase (up to 10^6 times) with rising humidity, thereby indicating the important role of humidity in DNA conduction[77]. In fact, participation of the DNA molecule in charge transport was confirmed by the high resistance of the environment, which exceeded up to 100 times the resistance of DNA itself[78]. It has been proposed that this phenomenon rests on the change of DNA permittivity, and therefore, the DNA conductivity due to adsorption of the water molecules on the DNA skeleton. However, a more elaborate study indicated[79] that humidity and electronic interactions between stacked base pairs change the electronic properties of DNA base pairs in such a way that the covalent structures of some nucleobases are converted to structures with separate charges (ionic type), where some covalent bonds are broken and the released electrons are weakly coupled to the lattice. Therefore, an applied electric field and elevated temperature can coax these electrons to contribute to DNA conductance.

When DNA is placed in the solvent the base atoms are capable of making hydrogen-bonded links to water molecules[80]. The hydration efficiency in this study is determined by the location of these atoms within an aromatic ring rather than on the DNA grooves. The structure of

---

[77]Humidity effects on the conductance of the assembly of DNA molecules, by D.H. Ha, et al., *Chem. Phys. Lett.* **355**, 405 (2002); Humidity dependence of charge transport through DNA revealed by silicon-based nanotweezers manipulation, by C. Yamahata, et al., *Biophys. J.* **94**, 63 (2008); Influence of humidity on the electrical conductivity of synthesized DNA film on nanogap electrode, by Y. Otsuka, et al., *Jpn. J. Appl. Phys.* **41**, 891 (2002).

[78]Dielectrophoresis of nanoscale double-stranded DNA and humidity effects on its electrical conductivity, by S. Tukkanen, et al., *Appl. Phys. Lett.* **87**, 183102 (2005).

[79]Water induced weakly bound electrons in DNA, by J. Berashevich and T. Chakraborty, *J. Chem. Phys.* **128**, 235101 (2008); How the surrounding water changes the electronic and magnetic properties of DNA, by J. Berashevich and T. Chakraborty, *J. Phys. Chem.* B **112**, 14083 (2008).

[80]Water molecules in DNA recognition II: a molecular dynamics view of the structure and hydration of the trp operator, by A.M. Bonvin, et al., *J. Mol. Biol.* **282**, 859 (1998).

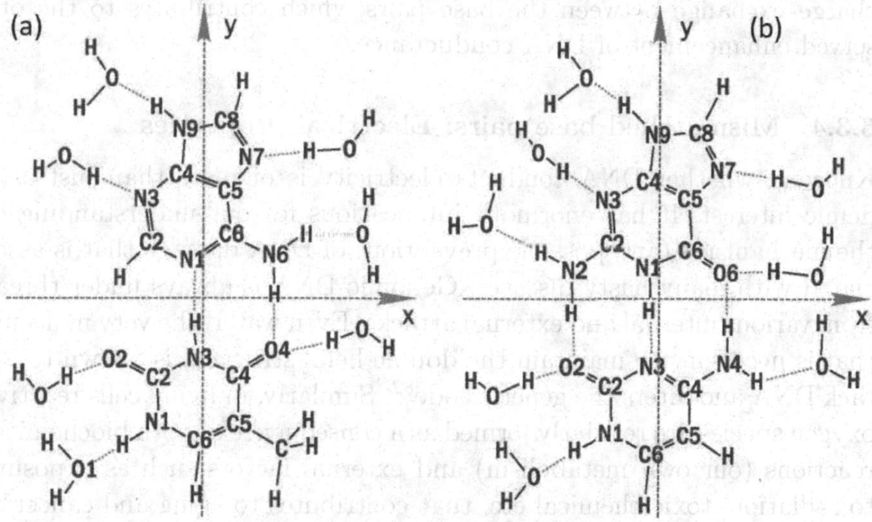

FIGURE 6.21
The structure of (a) water-(A:T) and (b) water-(G:C) geometries. The water molecules are attached by hydrogen bonds to the base pairs.

hydrated base pairs constructed by Berashevich and Chakraborty is shown in Fig. 6.21.

The interaction of molecular objects with water is known to be one of the important issues in molecular biology. In some cases these molecule-water interactions due to the charge exchange between them leads to a change in the electronic properties of the molecules and can result in ejection of an electron from the molecule to the solvent[81]. This ionic state usually exists because of the stabilization potential created by the surrounding solvent. For the DNA base pairs considered here, there are modifications of the HOMO-LUMO gap for the A:T and G:C pairs activated by the hydration. The electronic density distribution and the orbital phases are changed and the HOMO and LUMO energies are shifted. These lead to generation of weakly bound or even free charge carriers in the system. As a result, there is significant increase in the

---

[81] Charge transfer to solvent (CTTS) energies of small $X^-(H_2O)_{n=1-4}$ (X=F, Cl, Br, I) clusters: Ab initio study, by D. Majumdar, J. Kim, and K.S. Kim, *J. Chem. Phys.* **112**, 101 (2000); Structures, energetics, and spectra of electron-water clusters, $e^- - (H_2O)_{2-6}$ and $e^- - (HOD (D_2O)_{1-5}$, by H.M. Lee, S. Lee and K.S. Kim, *J. Chem. Phys.* **119**, 187 (2003).

charge exchange between the base pairs which contributes to the observed enhancement of DNA conductance.

6.3.4 Mismatched base pairs: Electrical properties

Knowing whether DNA conducts electricity is of more than just academic interest. It has enormous implications for our understanding of the mechanisms (and possible prevention) of DNA damage that is associated with many nasty diseases. Genomic DNA is always under threat from various internal and external attacks. Even water, the very molecule that is necessary to maintain the double helix structure is known to attack DNA and alters the genetic code[82]. Similarly, in living cells reactive oxygen species are regularly formed as a consequence of both biochemical reactions (our own metabolism) and external factors such as exposure to radiation, toxic chemical etc. that contributes to aging and cancer[83]. Throughout the life of any organism, maintaining the integrity of genetic material is of crucial importance. Given the amount of damage our DNA sustains on a daily basis, it is a small wonder that H. sapiens manages to walk around with a 4 billion base pairs genome! Protecting the genome from these damages is vital to the survival of a species.

During cell division the two single strands of DNA act as templates from which new complementary strands are created. This DNA replication process occasionally makes mistakes and if not detected in time will lead to genetic disease, or carcinogenesis. Interestingly, some of those mistakes in replication will produce a protein that works better than the original one and the mutant organism with the altered gene will spread through the population. The evolution actually depends on such positive mistakes. However, DNA has in its arsenal, an elaborate genomic maintenance apparatus to control the various DNA damages discussed above[84] and to preserve genome integrity. A clear understanding of the mismatch repair mechanisms essentially requires knowledge of the intrinsic properties of the mispairs, which distinguish them from the canonical Watson-Crick base pairs. Gathering knowledge of the complexity of

---

[82]How do DNA repair proteins locate damaged bases in the genome? by G.L. Verdine and S.D. Bruner, *Chemistry & Biology* **4**, 329 (1997).

[83]Cancer risk and oxidative DNA damage in man, by S. Loft and H.E. Poulsen, *J. Mol. Med.* **74**, 297 (1996); Mutational hot spots in DNA: Where biology meets physics, by J. Berashevich and T. Chakraborty, *Physics in Canada* **63**, 103 (2007).

[84]Recognition of DNA alterations by the mismatch repair system, by G. Marra and P. Schär, *Biochem. J.* **338**, 1 (1999.

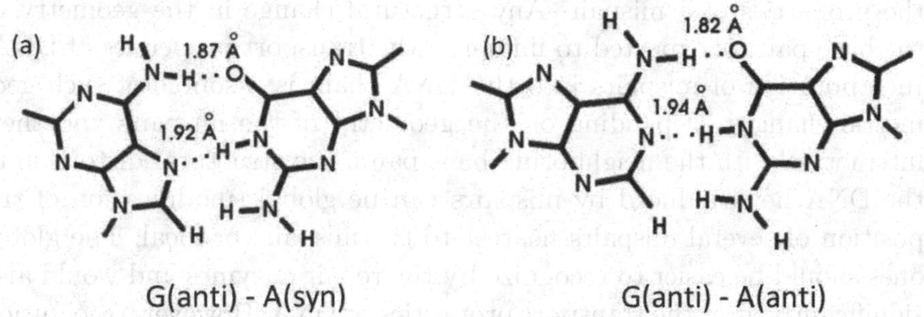

(a) H 1.87 Å
N–H··O
1.92 Å
N···H

G(anti) - A(syn)

(b) H 1.82 Å
N–H··O
1.94 Å
N···H–N

G(anti) - A(anti)

FIGURE 6.22
The geometries of the G (anti)·A(anti) and G (anti)·A(syn) conformations adopted by the G·A mispair (see Footnote 87).

recognition and repair of erroneous base pairing has been a major area of research in biology and chemistry[85]. Additionally, understanding the physical properties might help develop sensors capable of detecting the mismatched base pairs[86].

The presence of mispairs in DNA alters the thermodynamic properties of the molecule[87]. The mispairs also influence the magnetic properties. Although a mispair causes only local genomic changes in DNA, for small molecules the integral characteristics of DNA are modified by the mispair. As an example, changes in magnetic susceptibility of DNA due to the presence of a mispair are expected to occur for DNA containing up to 50 base pairs[88].

Charge transport through a finite-length DNA molecule connected to electrodes[89] containing a mispair[90] is a very promising route to study

[85]Mismatches and mutagenic lesions in nucleic acids, by T. Brown, *Aldrichimica Acta* **28**, 15 (1995).

[86]Single base mismatch detection by microsecond voltage pulses, by F. Fixe, et al., *Biosensors Bioelectronics* **21**, 888 (2005).

[87]Thermodynamics of G.A mispairs in DNA: Continuum electrostatic model, by J. Berashevich and T. Chakraborty, *J. Chem. Phys.* **130**, 015101 (2009).

[88]Unique magnetic signatures of mismatched base pairs in DNA, by V. Apalkov, J. Berashevich and T. Chakraborty, *J. Chem. Phys.* **132**, 085102 (2010).

[89]Charge transfer via a two-strand superexchange bridge in DNA, by X.F. Wang and T. Chakraborty, *Phys. Rev. Lett.* **97**, 106602 (2006).

[90]Quantum transport anomalies in DNA containing mispairs, by X.F. Wang, T. Chakraborty, and J. Berashevich, *Nanotechnology* **21**, 485101 (2010).

the properties of a mispair. Any structural change in the geometry of the base pairs is expected to influence the transport properties of DNA. Incorporation of mispairs into the DNA chain is a source of such geometric changes. Depending on the geometry of the mispairs and their interaction with the neighboring base pairs, the structural distortion of the DNA helix induced by mispairs can be global (modification of the position of several mispairs nearest to the mispair) or local. The global ones should be easier to recognize by the repair enzymes and would also significantly alter the transport properties of DNA. However, recognition of a mispair generating only local changes in the structure of DNA poses a major challenge. One example of a poorly recognized mispair is G·A, which is known to have several conformations. The two conformations G (anti)·A (anti) and G (anti)·A (syn) are the ones that cause only local changes in the DNA structure and are shown in Fig. 6.22.

For periodic poly(G)-poly(C) DNA sequence, a mispair inserted in the middle of the chain acts as a scatterer of the propagating charges. That scattering results in a suppression of the current through the DNA chain. The current versus the applied bias voltages, calculated for a tight-binding model[91] is shown in Fig. 6.23 without and with a mispair inserted in the middle of the chain. The I-V characteristic has a step-like behavior, where transitions between the steps occur when the chemical potentials of the contacts cross the HOMO or LUMO energy bands (that appear due to coupling between the nearest-neighbor base pairs). Clearly, the stronger suppression of the current is caused by the G(anti)·A(syn) mispair.

For a generic DNA molecule with random base-pair sequences, the effect of a mispair depend on the type of the mispair and the DNA sequence. If the current through DNA without a mispair is large then the corresponding electronic states of the molecule are well extended and provide good conductivity between the two contacts. The mispair in that case can be considered a point defect, which introduces an additional scattering and makes the extended states more localized, thereby suppressing the current. On the other hand, if the current through a DNA molecule is relatively small, then the electronic states are almost localized and introduction of a mispair can result in additional coupling between the quasilocalized states, making these states more extended. In this case the mispair increases the charge transport through the DNA.

---

[91]Electrical current through DNA containing mismatched base pairs, by N. Edirisinghe, et al., *Nanotechnology* **21**, 245101 (2010).

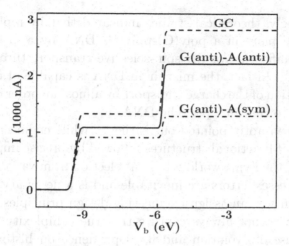

FIGURE 6.23
I-V characteristic of periodic poly(G)-poly(C) DNA molecule with no
mispair (dashed line), with the G(anti)·A(anti) mispair (solid line), and
G(anti)·A(syn) (dashed-dotted line).

Therefore, for a irregular DNA sequence, a mispair can be either (i) a
defect, which makes the extended electronic states more localized and
suppresses the current through the molecule, or (ii) a coupling element
between the quasilocalized states of the molecule, thus making those
states more extended and increases the current through the molecule.

A few other studies have also reported similar behaviors. Theoretical studies of transverse electron transport[92] through Watson-Crick and
non-Watson-Crick base pairs revealed that each base pair has a distinctive electrical signature determined by their electronic and chemical
properties[93]. Measurement of conductivities of individual DNA helices
chemically bound to two gold electrodes, using the STM break-junction
approach was also reported[94]. These measurements indicated that the
presence of a single base pair mismatch can be identified by the conductance of the molecule.

---

[92]Transverse tunneling current through guanine traps in DNA, by V. Apalkov and
T. Chakraborty, *Phys. Rev. B* **72**, 161102 (R) (2005).

[93]A DNA sensor for sequencing and mismatches based on electron transport
through Watson-Crick and non-Watson-Crick base pairs, by L.A. Jauregui and J.M.
Seminario, *IEEE Sensors J.* **8**, 803 (2008).

[94]Study of single-nucleotide polymorphisms by means of electrical conductance
measurements, by J. Hihath, et al., *PNAS* **102**, 16979 (2005).

Interestingly, theoretical studies indicated that by replacing one of the G:C base pairs in a poly(G)-poly(C) DNA by a G·A mispair, a profound modification of the spin-selective transport through DNA is also possible[95]. In fact, the mispair in DNA is capable of changing the spin polarization of the charge transport by almost an order of magnitude as compared to that in a periodic DNA.

As has been aptly pointed out earlier[96], while engineering relies on exact design of functional structures followed by precise implementation of the design, the living world relies on selection from variable processes. In those processes, errors are inevitable and is a necessary part of biology. Although we can assign engineering design principles to biological systems they do not always capture the true complexity of the living world. Because of evolution and its dependence on history, biological systems present a profoundly different paradigm from that of engineering. In designing future DNA based electronics, it would be wise to borne this neat aphorism in mind.

---

[95]Tunable spin-selective transport through DNA with mismatched base pairs, by V.M. Apalkov and T. Chakraborty, *J. Phys.: Condens. Matter* **26**, 475302 (2014).
[96]From DNA to transistors, by E. Braun and K. Keren, *Adv. Phys.* **53**, 441 (2004).

7

Epilogue and the road ahead

As Mark Twain (né Samuel Langhorne Clemens) once said[1], it is difficult to make predictions, particularly about the future. I would therefore refrain from such perilous actions and devote the rest of the pages to point to possible directions our nanoworld might proceed to face new challenges, all the while allowing us to make surprising discoveries, and a better understanding of nature at the nanoscale.

Dreams of blue: Roses have been cultivated for more than 5,000 years, and as many as 25,000 different species of roses currently exist. However, colors are traditionally limited to red, pink, yellow and white. But that is going to change soon. "Rose has been entwined with human culture and history. Blue rose in English signifies unattainable hope or an impossible mission as it does not exist naturally and is not breedable regardless of centuries of effort by gardeners. With the knowledge of genes and enzymes involved in flower pigmentation and modern genetic technologies, synthetic biologists have undertaken the challenge of producing blue rose by engineering the complicated vacuolar flavonoid pigmentation pathway and resulted in a mauve-colored rose". This is an excerpt from a recent publication[2], which perfectly embodies the uncurbed urge of humans to unravel Nature's every secret for its own pleasure (and, in modern days, for huge financial gain). The research on blue roses started in earnest in Japan in 1990, when they harvested the blue genes from petunias for this purpose and blue roses were first successfully created in 2004 by Suntory

---

[1] *To Light Such a Candle: Chapters in the History of Science and Technology*, by K.J. Laidler (Oxford U.P., New York 1998).

[2] Cloning and Expression of a Nonribosomal Peptide Synthetase to Generate Blue Rose, by A.N.N. Urs, et al., *ACS Synth. Biol.* **8**, 1698 (2019).

DOI: 10.1201/9781003090908-7

Global Innovation Center in Japan[3]. The mass marketing of true blue roses however, will have to wait for a while.

The nanoworld that we have visited in this book is strewn with the blue rose petals (not to mention fractal butterflies) where totally unexpected, and surprising phenomena are discovered in every turn of the way. Understanding these phenomena have significantly widened our knowledge of how nature works. Undoubtedly, there is much more that are still waiting to be explored. The discovery of the quantum Hall effect by von Klitzing did not close that chapter at all, but instead has inspired new theories and experimental discoveries far beyond the solid state electronics[4]. Similarly, the fractionally quantized Hall effect has given rise to such exotic phenomena as non-Abelian braiding statistics, quantum information processing, Majorana fermions, Topological insulators, etc.[5]. In fact, the quantum Hall effect and related discoveries have opened the door to limitless possibilities for exploration in the nanoworld of quantum materials, that have brought to the fore novel quantum systems. Quantum dots, quantum rings etc. are all part of those discoveries, where we expect to observe many more new and exciting physical phenomena and novel applications. Graphene and other Dirac-like materials are another front where there are novel effects waiting to be discovered. Then there are the electronic properties of DNA that are wide open for exploration. The electronic properties of DNA are now poorly understood, but it still has the potential to revolutionize the field of molecular electronics in the future. Progress in research on DNA conduction will be also important for a better understanding of the important field of DNA damage and its repair mechanisms. Finally, we should also mention the nascent field of *DNA computation*[6] that is poised to revolutionize the

[3]Blue roses – a pigment of our imagination? by T.A. Holton and Y. Tanaka, *Trends in Biotechnology* **12**, 40 (1994); Engineering of the Rose Flavonoid Biosynthetic Pathway Successfully Generated Blue-Hued Flowers Accumulating Delphinidin, by Y. Katsumoto, et al., *Plant Cell Physiol.* **48**, 1589 (2007); For a review, see, e.g., Recent advances in the research and development of blue flowers, by N. Noda, *Breeding Science* **68**, 79 (2018).

[4]40 years of the quantum Hall effect, by K. von Klitzing, T. Chakraborty, P. Kim, et al., *Nature Rev. Phys.* **2**, 397 (2020).

[5]A short introduction to topological quantum computation, by V.T. Lahtinen and J.K. Pachos, *SciPost Phys.* **3**, 021 (2017); Topological Insulators: Dirac equation in condensed matters, by S.-Q. Shen (Springer, New York 2012).

[6]*DNA Computing: New Computing Paradigms*, by G. Paun, G. Rozenberg and A. Salomaa (Springer, Heidelberg 1998); *Theoretical and Experimental DNA Computation*, by M. Amos (Springer, Heidelberg 2005).

art of computation in the quest to understand and simulate fundamental biological processes and algorithms occurring within cells.

Our nanoworld, as witnessed here is teeming with profound intellectual challenges and also provides ample opportunities for scientific revolutions. I believe that the topics discussed in this book can be considered as the magnifying lens to peer into many of the novel phenomena that are enriching the field of nanoscience with promises of fundamental and technical breakthroughs that are poised to benefit both the scientific community and society at large. After all, the traits that make human beings unique, will always remain the driving force to push farther the frontiers of nanoscale science for years to come. Until now, rapid growth of our insights into nature's workings have resulted in specializations and sprouting of various scientific disciplines that focus only on certain aspects of science. However, the nanoworld that we have visited here is highly interdisciplinary and involves important contributions from all branches of science and engineering. Understanding this complex world requires a multifaceted approach by scientists who will have to assume the role of the modern-day *uomo universale*, just as the pioneers, who centuries ago were the scientists, artisans and artists, all rolled into one.

A

Ten-martini proof

For the mathematically inclined, a sketch of the proof of the ten-martini problem (already introduced in Sec. 6.1.8), is presented below by Dr. S. Jitomirskaya[1].

We begin with some definitions. First, given the frequency α we introduce $\beta(\alpha)$, the exponential rate of approximation of α by the rationals

$$\beta(\alpha) := \limsup_{k \to \infty} -\frac{\ln ||k\alpha||_{\mathbb{R}/\mathbb{Z}}}{|k|} = \limsup_{n \to \infty} \frac{\ln q_{n+1}}{q_n} \in [0, \infty] \qquad (A.1)$$

where $||x||_{\mathbb{R}/\mathbb{Z}} = \inf_{\ell \in \mathbb{Z}} |x - \ell|$ and $\frac{p_n}{q_n}$ are the rational approximants of α (obtained by the continued fraction algorithm).

For Lebesgue almost all α, we have $\beta(\alpha) = 0$. Those α's are called Diophantine. However, for any given c, for topologically generic α we have $\beta(\alpha) > c$. Those α's are called Liouville.

We say phase θ is α-Diophantine if $\limsup_{k \to \infty} -\frac{\ln ||2\theta + k\alpha||_{\mathbb{R}/\mathbb{Z}}}{|k|} = 0$ and α-rational if $\theta = m + k\alpha$. Lebesgue almost all phases are α-Diophantine.

We will also need the transfer-matrix formalism. A formal solution of $H_{\lambda,\alpha,\theta} u = Eu$, $u \in \mathbb{C}^{\mathbb{Z}}$, satisfies the equation

$$\begin{pmatrix} E - 2\lambda \cos 2\pi(\theta + n\alpha) & -1 \\ 1 & 0 \end{pmatrix} \cdot \begin{pmatrix} u_n \\ u_{n-1} \end{pmatrix} = \begin{pmatrix} u_{n+1} \\ u_n \end{pmatrix}. \qquad (A.2)$$

Defining $S_{\lambda,E}(x) = \begin{pmatrix} E - 2\lambda \cos 2\pi x & -1 \\ 1 & 0 \end{pmatrix}$, it is clear that the behavior of solutions is governed by the behavior of the products $S_{\lambda,E}^n(\theta) := S_{\lambda,E}(\theta + (n-1)\alpha) \cdots S_{\lambda,E}(\theta)$. The Lyapunov exponent that measures

---

[1]Department of Mathematics, University of California, Irvine.

DOI: 10.1201/9781003090908-A

the exponential growth of transfer-matrices is given by

$$L(E) = \inf_n \frac{1}{n} \int_\Omega \ln \|S^n_{\lambda,E}(x)\| dx \text{ a.e. } x = \lim_{n\to\infty} \frac{1}{n} \ln \|S^n_{\lambda,E}(x)\| dx.$$
(A.3)

An important fact to note is that for the almost Mathieu operator the Lyapunov exponent is constant on the spectrum and can be explicitly determined: $L(E) = \max\{\ln|\lambda|, 0\}$ for $E \in \sigma(H_{\lambda,\alpha,\theta})$. This was conjectured by Aubry-Andre [AA][2] and later proved in [BJ][3].

The ten Martini challenge is to prove the Cantor spectrum for all $\lambda \neq 0$ and for all irrational α. One important special feature of the almost Mathieu family is the Aubry duality [AA]: a Fourier-type transform for which the family $\{H_{\lambda,\alpha,\theta}\}_\theta$ is a fixed point. One manifestation of the Aubry duality is that $\sigma(H_{\lambda,\alpha}) = \frac{1}{\lambda}\sigma(H_{1/\lambda,\alpha})$. This immediately reduces the problem to $|\lambda| \geq 1$. Moreover, there is a sharp transition at $|\lambda| = 1$, and for $|\lambda| > 1$, the Lyapunov exponents are positive on the spectrum, which is usually associated with Anderson localization (pure point spectrum with exponentially decaying eigenfunctions). This is important because of an argument by Puig [P][4], who offered a very simple route to the Cantor spectrum from the *arithmetic* localization. Puig's key observation was that the eigenvalues of the operator at a specific phase $\theta = 0$, are necessarily edges of open gaps. Therefore, if one can prove that those eigenvalues are dense in the spectrum (and they are if $H_{\lambda,\alpha,0}$ has pure point spectrum!), then the Cantor spectrum immediately follows.

Puig's argument however needs an additional requirement: that the corresponding eigenfuncations are exponentially decaying – i.e., precisely the Anderson localization statement. It is based on another manifestation of the Aubry duality: if $u_n \in \ell^2(\mathbb{Z})$ solves the eigenvalue equation $H_{\lambda,\alpha,\theta}u = Eu$, then $v^x_n := e^{2\pi i n\theta}\hat{u}(x + n\alpha)$ solves

$$H_{\frac{1}{\lambda},\alpha,x}v^x = \frac{E}{\lambda}v^x$$
(A.4)

for a.e. x, where $\hat{u}(x) = \sum e^{2\pi i n x} u_n$ is the Fourier transform of u.

---

[2]Analyticity breaking and Anderson localization in incommensurate lattices, by S. Aubry, G. André, *Ann. Israel Phys. Soc.* **3**, 133 (1980) [AA].

[3]Continuity of the Lyapunov exponent for quasiperiodic operators with analytic potential by J. Bourgain and S. Jitomirskaya, *J. Stat. Phys.* **108**, 1203 (2002) [BJ].

[4]Cantor spectrum for the almost Mathieu operator, by J. Puig, *Comm. Math. Phys.* **244**, 297 (2004) [P].

If $\theta = 0$, this leads to a quasiperiodic solution to the dual eigenvalue problem at $\frac{E}{\lambda}$, which makes the corresponding transfer-matrix reducible to an upper-triangular matrix with the 1s on the diagonal. If $\beta(\alpha) = 0$, this matrix can be further reduced to a constant matrix by solving the cohomological equation. Finally, the latter cannot be the identity matrix, because otherwise the original problem would have two linearly independent decaying solutions at E, which is impossible by the Wronskian considerations. Therefore the matrix is parabolic, and a well-known fact of Floquet theory implies that $\frac{E}{\lambda}$ is then an edge of a non-collapsed gap (indeed it can be checked directly that if a matrix is reducible to a constant parabolic, then adding a sufficiently small constant of appropriate sign to the upper left entry results in a uniformly hyperbolic cocycle, the corresponding energy then being in a gap). The argument applies equally well to all α-rational phases θ. Moreover, this argument also proves that the E is a gap edge whenever the corresponding transfer-matrix is reducible to an upper-triangular matrix with the 1s on the diagonal, which is true whenever it is reducible to a constant matrix and the integrated density of states $N(E)$ takes values in $\alpha\mathbb{Z} + \mathbb{Z}$.

Application of this argument therefore boils down to establishing that one has a dense set of eigenvalues with exponentially decaying eigenfunctions for $\theta = 0$. This was done in [JKS][5] building on the *arithmetic* localization argument developed in [J][6] (with some ideas going back to [NP][7]) for all Diophantine α. More precisely, [J] established the localization for all α-Diophantine θ, and in [JKS] the argument was extended to all α-rational θ.

Equipped with this arithmetic localization result, Puig's argument established the Cantor spectrum for all α with $\beta(\alpha) = 0$, thus a.e. α.

The arithmetically opposite case of the Liouville α (large $\beta(\alpha)$) was also known since the work of Choi, Eliott, Yui [CEY][8]. For Liouville α, there is a sequence of rationals p_n/q_n such that $|\alpha - p_n/q_n| < e^{-cq_n}, c < \beta(\alpha)$. Therefore, if we know that the gaps in the spectrum of H_{p_n/q_n}

[5]Continuity of the Lyapunov exponent for analytic quasiperiodic cocycles, by S. Jitomirskaya, D. Koslover, and M. Schulteis, *Erg. Theory Dyn. Syst.* **29**, 1881 (2009) [JKS].

[6]Metal-insulator transition for the almost Mathieu operator, by S. Jitomirskaya, *Annals of Math.* **150**, 1159 (1999) [J].

[7]Anderson localization for the almost Mathieu equation: A nonperturbative proof, by S. Jitomirskaya, *Comm. Math. Phys.* **165**, 49 (1994) [NP].

[8]Gauss polynomials and the rotation algebra, by M.D. Choi, G.A. Eliott, and N. Yui, *Invent. Math.* **99**, 225 (1990) [CEY].

are larger than e^{-cq_n}, then we know that those gaps will persist, by continuity, in the spectrum of H_α. For $\alpha = \frac{p}{q}$ the spectrum of $H_{\alpha,\theta}$ can have at most $q - 1$ gaps. It appears that all those gaps are open, except for the middle one for even q [CEY]. In fact, Choi et al. [CEY] obtained an exponential lower bound on the size of the individual gaps which immediately implied the presence of the Cantor spectrum for α with sufficiently large $\beta(\alpha)$ [CEY].

Can one make these two approaches meet? The "follow the rational gaps" approach of [CEY] consisted of establishing the lower bounds of gap sizes for rational α and proceeding by continuity. Its limits are governed by two aspects: (i) how sharp the lower bounds on sizes of the gaps for $\alpha = \frac{p}{q}$ can be made, and (ii) the quantitative continuity of the spectra: how far can the spectrum of H_α venture away from that of H_β with respect to $|\alpha - \beta|$.

To address the first issue, we have proven in [AJ] that if $\frac{p}{q}$ is close enough to α, then all open gaps of $\Sigma_{\lambda,\frac{p}{q}}$ have the size of at least $e^{-(|\ln\lambda|+\epsilon)q/2}$, for any $\epsilon > 0$. Here, $\ln\lambda$ is the Lyapunov exponent, and for various reasons this result is optimal. The quantitative continuity of the spectra cannot possibly be better than Lipschitz continuity, but even that is unrealistic. The available result is Hölder $\frac{1}{2}$, meaning that given E in the spectrum of H_β there is an E' in the spectrum of H_α with $|E - E'| < C|\alpha - \beta|^{1/2}$ [AMS][9]. Together, this leads to the Cantor spectrum result (in fact, a stronger, "dry ten martini" statement) for $e^{-\beta(\alpha)} < |\lambda| < e^{\beta(\alpha)}$.

However, we can do better than that! Arguing by contradiction and assuming that there is an interval in the spectrum, we obtain that for E in that interval there is a change of coordinates $B_E(x) : \mathbb{R}/\mathbb{Z} \to \mathrm{SL}(2,\mathbb{R})$, analytic in (x, E), that reduces the transfer matrix cocycle $S_{\lambda,E}(x)$ to a cocycle of rotations:

$$B_E(x + \alpha) \cdot S_{\lambda,E}(x) \cdot B(x)^{-1} \in \mathrm{SO}(2,\mathbb{R}). \qquad (A.5)$$

The regularity of rotations allows us to conclude the Lipschitz continuity for the spectrum, leading as above, to the Cantor spectrum for an enlarged range $e^{-2\beta(\alpha)} < |\lambda| < e^{2\beta(\alpha)}$. This, however, is the absolute best one can do from the Liouville side, and the rest should come from the localization side.

---

[9]On the measure of the spectrum for the almost Mathieu operator, by J. Avron, P. van Mouche, and B. Simon, *Commun. Math. Phys.* **132**, 103 (1990) [AMS].

For how large values of $\beta(\alpha)$ can the localization hold? That was the subject of a conjecture in [CONGR][10] that there is a second sharp transition: (i) for α-Diophantine θ, $H_{\lambda,\alpha,\theta}$ has the Anderson localization if $|\lambda| > e^{\beta(\alpha)}$, and, (ii) for all θ, $H_{\lambda,\alpha,\theta}$ has purely singular continuous spectrum (hence no eigenfunctions) if $1 < |\lambda| < e^{\beta(\alpha)}$. It was eventually proved in [JL1][11] (see also a dual result in [JL2][12]), with a prior non-arithmetic result in [AYZ][13]. The first part was conjectured but was unknown at the time [AJ][14] was written, but the second part was already clear. α-Diophantine phases θ are, in some sense, the best for the localization purposes, but initially we expected the same localization threshold for $\theta = 0$ that was needed for Puig's argument described above. It was clear that it would be difficult, but there was some room to spare since our "proof-by-contradiction" trick on the Liouville side increased the range to $|\lambda| < e^{2\beta(\alpha)}$, and so a localization result for $\theta = 0$, $|\lambda| > e^{c\beta(\alpha)}$, and for any $c < 2$ would have been sufficient. However, in the process of trying to prove this, we realized that for α-rational θ, the expected transition point should be modified to be not $|\lambda| = e^{\beta(\alpha)}$ but $|\lambda| = e^{2\beta(\alpha)}$, because resonances "double up": the same lattice points n become both frequency-resonant (small $||n\alpha||$) and phase-resonant (small $||2\theta + n\alpha||$).

We conjectured it in [AJ], and it is still unresolved with the best result being reported in [LIU][15]. If that conjecture is true then the method based on localization at $\theta = 0$ could only work for $|\lambda| > e^{2\beta(\alpha)}$, but still leaves the set of α with $|\lambda| = e^{2\beta(\alpha)}$ untreated (even setting aside the need to deal with the cohomological equation).

Clearly, something else had to be developed, at least to cover α with $|\lambda| = e^{2\beta(\alpha)}$. Our localization-side proof covers, in fact, all the α such that the Anderson localization holds for a.e. θ. With the current state

---

[10] Almost everything about the almost Mathieu operator, II., by S. Jitomirskaya, Proc. of XI Int. Congr. of Math. Phys., *Int. Press* **373** (1995) [CONGR].

[11] Universal hierarchical structure of qp eigenfunctions, by S. Jitomirskaya and W. Liu, *Annals. Math.* **187**, 721 (2018) [JL1].

[12] Universal reflective-hierarchical structure of quasiperiodic eigenfunctions and sharp spectral transitions in phase, by S. Jitomirskaya, and W. Liu, arxiv:1802.00781 [JL2].

[13] Sharp phase transitions for the almost Mathieu operator, by A. Avila, J. You, and Q. Zhou, *Duke Math. J.* **166**, 2697 (2017)[AYZ].

[14] The ten-martini problem, by A. Avila and S. Jitomirskaya, *Ann. Math.* **170**, 303 (2009) [AJ].

[15] Almost Mathieu operators with completely resonant phases, by W. Liu, *Ergodic Theory Dynam. Syst.* **40**, 1875 (2020) [LIU].

of knowledge, that would mean all α with $|\lambda| > e^{\beta(\alpha)}$. However, this was unknown when we were working on [AJ], and the most technically involved and complicated part of the paper was establishing such a localization result which we were able to do for α with $|\lambda| > e^{\frac{16}{9}\beta(\alpha)}$. This was sufficient to complete the argument since $\frac{16}{9} < 2$.

The localization proof stems from the method presented in [J], but it still required novel and rather complicated ways of dealing with exponentially strong resonances. Assuming this localization result, the argument is based again on the Aubry duality. If E is an eigenvalue of $H_{\frac{1}{\lambda},\alpha,\theta}$ with exponentially decaying eigenfunction u_n and θ is not α-rational, then $v_n^x = e^{2\pi i n \theta}\hat{u}(x + n\alpha)$ is not real. Then by (A.4) both v_n and $\bar{v}_n$ solve the dual eigenvalue equation. They can be shown to be linearly independent, and this can be used to create a change of coordinates $B_E(x) : \mathbb{R}/\mathbb{Z} \to \mathrm{SL}(2,\mathbb{R})$ reducing $S_{\lambda,E}(x)$ to a constant rotation matrix: $B_E(x+\alpha) \cdot S_{\lambda,E}(x) \cdot B(x)^{-1}$ is constant. It is also easy to see that in this case the integrated density of states $N(E)$ must take a value of the form $\pm\theta + k\alpha$. Cocycles for which such a change of coordinate exists are called reducible. Let Λ be the set of E such that $S_{\lambda,E}(x)$ is reducible. Here again comes an argument by contradiction. Assume that there is an interval J in the spectrum. Then, by (A.5) we have an analytic conjugacy to a (possibly non-constant) rotation matrix for all $E \in J$. Thus a question of reducibility reduces to the solution of a standard cohomological equation. This solution is easy under an assumption $\beta(\alpha) = 0$ (taking care of this case, in particular, for $|\lambda| = 1$, since no localization is needed up to this point).

What we have proved for the $\beta(\alpha) > 0$ regime using further, a bit more intricate, analytic continuation ideas, is as follows. If $\beta(\alpha) < \infty$ then, under an assumption of J in the spectrum, if the set $\Lambda \cap J$ has positive measure, then it necessarily contains an interval contained in J. However, then it would contain E with $N(E) \in \alpha\mathbb{Z} + \mathbb{Z}$, and those, by Puig's argument, are gap edges, a contradiction. Hence $\Lambda \cap J$ must be of measure zero. But one can also show that the integrated density of states $N(E)$ is a nonconstant analytic function on J. It implies that the set of θ for which for some $E \in \Lambda$, the integrated density of states $N(E)$ takes a value of the form $\pm\theta + k\alpha$ is also of measure zero. This contradicts the a.e. localization result for a.e. θ for $|\frac{1}{\lambda}| > e^{\frac{16}{9}\beta(\alpha)}$, thus completing the proof.

We should note that the proof contains also a result for $|\lambda| = 1$ (the case $\beta(\alpha) > 0$ was taken care of from the Liouville side). It was, however, already known at the time [AJ] was written, through a combination of [AK][16] and [ONE][17]. Recently a very simple proof was found in [JK][18]).

[16]Reducibility or nonuniform hyperbolicity for quasiperiodic Schrödinger cocycles, by A. Avila and R. Krikorian, *Ann. Math.* **164**, 911 (2006) [AK].

[17]Zero measure of the spectrum for the almost Mathieu operator, by Y. Last, *Comm. Math. Phys.* **164**, 421 (1994) [ONE].

[18]Critical almost Mathieu operator: hidden singularity, gap continuity, and the Hausdorff dimension of the spectrum, by S. Jitomirskaya and I. Krasovsky, arxiv:1909.044290 [JK].

Index

Almost Mathieu equation, 190
Anisotropic mass tensor, 42
Anyons, 35
Arrow of time, 194
Artificial atoms, 47, 52

base pairing, 206
Biased bilayer graphene, 138
Bilayer graphene, 132
Black phosphorous, 159
Blue rose, 217
Bose statistics, 33
Butterflies
    in bilayer graphene, 178
    interacting electrons, 181
    in monolayer graphene, 171

Cantor set, 188
    self-similar, 190
Chirality, 128
Classical plasma, 27
Compressibility, 31
Confinement
    Dirac electrons, 149
    Dirac fermions, 148
    Electron, 11
Confinement potential
    Dirac fermions, 148
    Graphene quantum ring, 150
Cryptography, 70

Demon exorcism, 199
Demon in a quantum ring, 201
Demon in quantum dots, 199

Dirac Hamiltonian, 127, 128
Dirac points, 129
Dirac-like materials, 158
DNA
    base pairs of, 206
    bases of, 206
    Discovery, 206
    Double helix, 206
    hereditary material, 206
    Structure and function, 206
DNA (deoxyribonucleic acid), 205
DNA conduction, 209
    Humidity assisted, 210
DNA damage, 212
DNA double helix structure, 206
DNA electronics, 209

Emission spectra, blue shift, 84
Entropy, 194
Entropy and disorder, 194
Exchange hole, 30

Fermi energy, 12
Fermi statistics, 33
Filled Landau level, 28
Filling factor, 18, 19, 26
Flatband condition, 171, 175
Flux quantum, 19
Fock-Darwin level, 50, 52
FQHE in graphene, 132
Fractal butterflies, 165
Fractional electron charge, 30
Fractional quantum Hall effect, 23

Fractional quantum Hall states
    in bilayer graphene, 136
    in butterfly spectrum, 186
    in graphene, 132
    Spin configurations, 37
    Spin-reversed, 37
Fractional statistics, 35
Fractionally charged
    quasiparticles, 31, 32

Galilean metric, 41
Galileo's scaling law, 4
Germanene, 158
Graphene
    crystal structure, 121
    molecular adsorption, 156
Graphite
    band structure, 122
    magnetic energy levels, 124
Graphite structure, 120

Hall conductance quantization, at
    half-integer, 131
Hall plateaus, 20
Hall resistance, 16
Harmonic oscillator, 53
Harper equation, 166, 167, 190
Hartree-Fock energy, 30
Heat death of the universe, 195
Helicity, 128
Heterostructures, 11, 15
    MgZnO/ZnO, 43
Hofstadter butterfly, 166
    Energy spectra, 175

Incompressible liquid, 30, 31
Incompressible state, 29, 30
Incompressiblr state, in a periodic
    potential, 185
Interacting electrons, in graphene,
    147

Intersubband-cyclotron,
    transition, 81

Key distribution problem, 71

Landau level
    degeneracy, 18
    index, 129
Landau levels, 18, 49
    bilayer graphene, 135
    in graphene, 129
Laughlin state, 39
    geometry fluctuation, 39
    origin of incompressibility, 39
    in ZnO heterostructure, 43
Laughlin's theory, 25
Laughlin's wave function, 26
Linear energy dispersion, 124

Magic number, 56
Mass anisotropy, 40
Massive chiral fermions, 135
Massive Dirac fermions, 135
Massless Dirac fermions, 127
Maxwell's demon, 195, 196
Mismatched base pairs, 212
Modular arithmetic, 71
Moiré structure, 176
Molecular fingerprinting, 77
MOSFET, 11, 14
Multiple quantum rings, 104

Nanometer, 9

One-component plasma, 27, 29
Optical absorption, 151
Optical transitions
    between electronic subbands,
    78
    Elliptical quantum dots, 69
    Graphene quantum dots, 149

Intersubband-cyclotron, 82
Quantum cascade laser, 81
Quantum dot, 63
Quantum rings, 111, 115

Pair-correlation function, 25
Periodic potential, 169
Persistent current, 88, 89, 105
    Graphene quantum ring, 150,
        151
Phosphorene, 160
Photon correlation, 75
Plasma analogy, 32
Poorly recognized mispair, 214
Pseudospin, 127

Quantized Hall effect, 17, 18
    fractionally-, 24
Quantized Hall plateau, 18
Quantum cascade laser, 78
    in a magnetic field, 79
Quantum confinement, 10, 11
Quantum cryptography, 72
Quantum dot cascade laser, 83
Quantum dot hydrogen, 64
Quantum dots, 47
    Anisotropic, 68
    in graphene, 148
    spin textures, 64
Quantum Hall effect, 16
    in graphene, 130
    Half-integer, 131
    Integer, 20
Quantum key distribution, 72
Quantum ring, electron spin,
        100
    magnetization, 92
    Non-reciprocal, 201
    Rashba spin-orbit interaction,
        105

topological charge, 107
    in ZnO, 110
Quantum rings
    in graphene, 150
    interacting electrons, 91
    optical spectroscopy, 94
    Spin demon, 201
Quasiparticles and quasiholes, 31

Rashba spin-orbit coupling, 201

Scotch-tape method, 125
Second law of thermodynamics,
        193
Self-similar pattern, 177
Silicene, 158
Single-photon detector, 73
Single-photon source, 75
Sorting demon of Maxwell,
        195
Spin textures, 108
Spin-orbit coupling, 58
    constant, 61
Spin-orbit interaction, 60
    Energy dispersion, 61
    Energy splitting, 63
    Hamiltonian, 61
    in heterostructures, 61
Spin-unpolarized wave function,
        36
Spinor wave functions, 127
Statistical parameter, 35
Subbands, 12
Suppression of current due to
        mispairs, 214

Ten-martini problem, 190, 191,
        221
Ten-martini proof, 221
Thermodynamic time asymmetry,
        194

Tilted field
  bilayer graphene, 144
  emission spectra, 80
  mass anisotropy, 42
Tilted-field experiment, 37
Topological charge, 65, 107
Trapping potential, 148
Two-dimensional electron systems,
  11

Unimodular metric, 41

Vandermonde determinant, 26, 29
von Klitzing constant, 21, 22

Winding number, 64, 107

Zero field spin splitting, 60
ZnO quantum ring, 111
  optical transitions, 111

Printed in the United States
by Baker & Taylor Publisher Services